Energy for Development

Also by Douglas F. Barnes

Cleaner Hearths, Better Homes: Improved Stoves for India and the Developing World

The Challenge of Rural Electrification: Strategies for Developing countries.

The Urban Household Energy Transition: Social and Environmental Impacts in the Developing World

Rural Energy And Development: Improving Energy Supplies for Two Billion People

Environmental Health and Traditional Fuel Use in Guatemala

Restoring Balance: Bangladesh's Rural Energy Realities

ELECTRIC POWER FOR RURAL GROWTH

ELECTRIC POWER
FOR RURAL GROWTH:

HOW ELECTRICITY AFFECTS RURAL LIFE
IN DEVELOPING COUNTRIES

Second Edition

Douglas F. Barnes

Routledge
Taylor & Francis Group

LONDON AND NEW YORK

First published 2014 by Westview Press

Published 2018 by Routledge
52 Vanderbilt Avenue, New York, NY 10017
2 Park Square, Milton Park, Abingdon, Oxon OX14 4RN

Routledge is an imprint of the Taylor & Francis Group, an informa business

Library of Congress Cataloging-in-Publication Data

Barnes, Douglas F.
Electric Power for Rural Growth: How Electricity Affects Rural Life in Developing Countries / Douglas F. Barnes.
2014

ISBN 13: 978-0-367-01223-6 (hbk)
ISBN 13: 978-0-367-16210-8 (pbk)

Library of Congress Control Number: 2014919639

Dedication

To my family, my wife Mary Ann, my daughter Andrea and my son Chris, who endured my time away for the frequent international travel that I spent writing the first edition of this book.

Contents

Preface

This is the second edition of my first book, published in 1988 based on research during my first job at Resources for the Future. At the time, international donors were having serious doubt about the benefits of rural electrification for developing countries. Today several aspects of this debate have changed. The international donors now are solidly behind the idea that poor people should have access to electricity. With the United Nations Sustainable Energy for All program highlighting the importance of providing electricity to even the most remote corners of the earth, the debate over the efficacy of rural electrification is historically relevant, but over.

In the 1980s, no one was counting the number of people that were without electricity, but projecting backward from recent trends, I estimate that during those times about 2.6 billion people had no access to electricity from grid systems. Today that number has shrunk to just over 1 billion, with about 2 billion people having gained access during that 30-year period. Countries like China, Thailand, Brazil and Mexico now have electricity access rates that are well over 95 percent, and they are working on the last remaining pockets of people without electricity. But this does not mean that the task is complete. Even a large country like India still has more than 200 million people without electricity, and in rural Africa, astoundingly, only 1 in 8 people in rural areas has electricity. So complacency is not the order of the day, and actions are still needed to bring a modern life to those living in extreme poverty or very remote areas.

The definition of rural electrification also has evolved due to technological changes in the way electricity service can be provided in more remote areas. The definition is becoming increasingly more complicated as new devices and systems are developed that offer a variety of electricity service levels.

With the exception of flashlights, historically the main source of electricity came from national or local grid systems. In addition to national grids, some regions had local or isolated grid systems generally based on either micro-hydro or diesel generators, and the distribution system did not extend beyond a local area, such as a town or group of towns. In the 1990s, the first generation of household off-grid technologies, such as solar home systems, was developed and marketed to consumers. These systems involved photovoltaic panels, coupled with a battery, mostly for a single home. More recently, even smaller household lighting systems have been developed. Although often unrecognized, for many years rural people have used car or motorcycle batteries as a source of power for watching television and lighting their houses. Today there are even more new developments, such as thermo-electric devices that can turn heat into small amounts of electricity when attached to household cooking stoves. These technologies are still in the early stages of marketing and development.

Among this wide variety of new systems and devices, the cost and level of service differ markedly. Moving from the grid to off-grid technologies increases the cost per kilowatt-hour and decreases the service. When it is reliable, the grid is the most desirable electricity service, but there are well-known constraints to extending grid electricity to remote areas of developing countries. These include the high costs involved in reaching remote areas and lack of local capacity to use much electricity. In such situations, off-grid technologies often are less expensive than the grid.

Since the original writing of this book, the development impact of grid electricity on rural households has been the subject of a significant amount of research. I am glad to report that the findings of this early study have been validated for the most part. Unfortunately, few current studies focus on the impact of solar home systems or small lighting systems on socioeconomic development. It is well-known that certain activities cannot be accomplished by relying on the low power levels available through solar home or smaller photovoltaic systems. The question is whether this matters or should such technologies be considered "pre-electrification"—important in their own right but awaiting further expansion of grid electricity systems. These important new questions can only be answered by new research. In the meantime, I offer this second edition of my impact study of rural electrification that was conducted in the 1980s. The purposes of this book are to inform the issues in the public policy debate, advance empirical knowledge about the major issues and reach conclusions on the efficacy of various ways to implement rural electrification for development.

This entirely new production of the original book offers important historical information on the state of rural electrification in the 1980s. I have updated the text and titles, and the tables and charts have been revised for clarity. Some material that is no longer relevant has been omitted. I also have added a new chapter that summarizes the development of benefit evaluation methods, along with findings from recent research on the impact of rural electrification for development. But overall, the issues identified in the 1980s remain extremely relevant today in the context of the new international emphasis on providing modern energy access for all.

During the 1980s, the belief of rural electrification as a catalyst for socioeconomic change in developing countries had come under fire. The research in this book came about as a reaction to the public policy debate over rural electrification for development. In the context of the polemics, most analysts agreed that there was a great need to understand exactly how rural electrification affected the lives of rural people. The 1980s study was designed to go beyond a review of the issues in the debate over rural electrification to include detailed empirical analyses of the socioeconomic impact of rural electrification. It was believed that adding new information to the public policy debate was more important than recapitulating the competing positions. The project proved challenging because it cut across the boundaries of several disciplines, including economics, sociology and population research.

The study began in 1980 with funding from the Office of Energy of the United States Agency for International Development (USAID) and was completed with assistance from Resources for the Future (RFF) and the World Bank. However, the views expressed in this book are those of the author and do not necessarily reflect those of the organizations that supported the study.

During the original study, Joy Dunkerley, Bill Ramsay and Lincoln Gordon offered their continual support. Without their efforts and backing, this work would never have been completed. Lalit Sen gave advice and support for all phases of the work. Henry Peskin provided comments on draft versions of various chapters in the original manuscript and later worked with me on new studies that further refined both the methodology and our understanding of the benefits of rural electrification. I thank Bob Ichord for his many suggestions both early and late in the project. Finally, I wish to thank Karl Jechoutek and Jim Fish of the World Bank for their encouragement.

The comparative nature of the research was made possible by reports written expressly for the project on India, Colombia and Indonesia. B. B. Samanta and A. K. Sundaram prepared an excellent report on the socioeconomic impact of rural electrification in India. A benefit-cost report for India was ably completed by R. Venkatesan, along with K. Ravi Shankar, Sunil Bassi and R. K. Pachauri. Eduardo Velez and Janice Brodman each contributed valuable reports on the socioeconomic impact of rural electrification in Colombia and Indonesia, respectively. The first edition of the book was dedicated to Frederick Fliegel, an influential teacher at the University of Illinois who passed away just prior to its publication.

Finally, I have worked with Norma Adams, the editor of this second edition for more than 20 years. She edited the final version of this book with her usual diligence and flair for language. I also owe her a debt of gratitude for her advice on the final version.

Douglas F. Barnes
2014

1. Historical Public Policy Controversy

The blind faith placed in rural electrification during the 1960s and early 1970s as being a key to energy development policy bears resemblance to more recent calls by international organizations of Sustainable Energy for All (UN 2012). During those times, rural electrification was perceived almost as a magical force that would transform poor areas into highly productive regions. Advancing power lines into poor rural areas was synonymous with providing the necessary infrastructure for quickly bringing them into the 20th century. Communication, lighting, productivity increases, reduction in birth rates, the elimination of traditional customs blocking modernization and many other benefits would flow from a reliable supply of electricity to rural areas.

Electricity obviously plays an indispensable role in modern society and modern life; that a society could progress without a substantial commitment to rural electrification seemed almost incomprehensible to early planners, as it does even today. Certainly, electricity is a prerequisite for attaining the level of productivity and quality of life experienced in developed countries. However, this historical optimism contained within it the seeds of overselling a good policy, which perhaps provides cautionary lessons for today's energy policy makers.

If all these positive aspects concerning rural electrification were true, then why was there a controversy? Why were there questions about the priority rural electrification should have in development? The early optimism was clouded by reports in the 1980s from a number of countries indicating that the anticipated development effect of rural electrification had been slow to materialize. Prospective customers in villages and communities with new electricity service did not adopt it at the rate envisioned. When fewer people in rural communities took advantage of available electrical service, there was less chance that electrification programs would have an impact on rural productivity and quality of life. And since most rural electrification projects in developing countries were subsidized, lagging demand for electricity may have translated into a substantial financial strain on the power distribution companies, whether public or private. Also, concerns were raised regarding the equity of rural electrification investments and subsidies. Since mainly better-off villagers could afford to adopt electrification initially, there was the risk of worsening the gap between rich and poor.

The relevance of the historical debate for today is that the criticisms of rural electrification and lessons learned from poorly implemented past programs seem to have been forgotten. The required subsidies for reaching the poorest people in the world's most remote locations can be quite high. The agencies promoting electricity to those without it need to be financially sustainable to provide the type of service that can have significant socioeconomic impacts. For new advocates of rural electrification, the historical debate offers forgotten lessons.

The Historical Debate

Energy policy for developing nations before 1975 basically was rural electrification policy. Before the early 1970s, interest in rural electrification for developing nations had been stimulated by the tremendously successful programs implemented in the United States. The creation of the Rural Electrification Administration (REA) in the mid-1930s had led to enormous growth in rural electrification, a burst of rural productivity and dramatic improvement in the quality of rural life in the U.S. Rural electrification was administered through a separate government agency providing subsidized loans to rural cooperatives. These cooperatives constructed the distribution systems and administered the consumer billing and collection activities, but were seldom involved in the actual generation of electricity. The cooperatives generally were consumer-oriented companies with the political ideals that electrical energy should be available to consumers at reasonable rates. Based on the experience of the cooperatives, the U.S. Agency for International Development (USAID) became the first donor agency with major involvement in rural electrification.

In the early U.S. programs, the rural electric cooperatives and commercial power companies were in conflict with one another. The early literature published by the REA contains many accounts of public power companies "skimming the cream." That is, the commercial companies would attempt to connect the most densely populated rural communities, leaving the expensive scattered consumers to the cooperatives. The advantage of establishing rural electric cooperatives was that they advocated the adoption of electricity by all rural consumers, and also promoted the purchase of appliances available in urban areas. The cooperatives had an institutional interest in connecting all rural consumers rather than just those most profitable to them.

Through the cooperatives, the REA was quite successful in extending electricity to rural areas. But owing to its success, many early problems encountered by the rural electrification programs have been forgotten. For instance, Muller (1944) assessed the state of U.S. rural electrification as follows:

> Generally speaking, three factors influence the rural use of energy. There is little of the high industrial load and none of the dense residential load, which have made utility operation profitable in urban areas. Second, farmers and other rural residents are new, naturally conservative consumers. The purchase of electric energy becomes for most of

them a considered alternative to other possible uses of their income. Rural consumers generally are still in the stage of exploring the economy and convenience of electric light and power, and they still hesitate to use these facilities in large quantities. Rural communities have not come to the realization of the advantages of electrification with respect to local industries. The third factor is the economic status of most rural residents. They have not found it easy to pay for wiring, equipment, or energy. In 1935, ...42.1 percent of the country's farms were operated by tenants; 34.5 percent were mortgaged, rendering it somewhat difficult for operators to make extensive improvements. Finally, of course, a fairly high percentage of farms are submarginal and not profitable even if electrified. These factors have worked together to keep rural loads and consumption per consumer low. (Frederick William Muller, *Public Rural Electrification* [Washington, DC: American Council on Public Affairs, 1944], paraphrased)

These same issues concerning rural electrification have been raised in developing countries today. Farmers in most developing countries lack the income or purchasing power of U.S. farmers of the 1930s and 1940s, yet the expectation for developing countries was framed on the U.S. experience. The early projects in such countries as Colombia, Ecuador, and the Philippines were expected to promote development like that which had occurred in the U.S. in the 1940s and 1950s. However, these early expectations were not realized.

The oil crisis in 1974 was followed by sharp increases in the price of commercial energy, which led to increased energy research and a movement to examine alternative forms of energy. The unrealistic expectations for rural electrification, along with increased emphasis on alternative forms of energy, led to a debate over rural electrification policy in the 1980s.

The reports on rural electrification generally can be classified into three time periods spanning from the early 1970s to the 1980s. In the early 1970s, several reports were favorable to rural electrification, giving optimistic assessments of the socioeconomic impacts. In the mid-1970s, a stream of reports criticized the then current rural electrification policy. As mentioned above, these reports were in response to the growing concern over energy and dissatisfaction with rural electrification as the major energy policy for developing nations. Finally, the early 1980s witnessed a series of new examinations of rural electrification to determine the actual benefits for developing nations (Smith, Mehta, and Hayes 1983; USAID 1983; IDB 1979a). Since 2000, a new line of research has more critically examined the direct and indirect benefits of rural electrification (chapter 9).

Early Favorable Studies

The early studies of rural electrification highlighted the favorable consequences for socioeconomic development. Studies in Colombia and the Philippines offered glowing accounts of the benefits of rural electrification. The Colombia study indicated that rural electrification could act as a catalyst to development (Ross 1972). The study found household consumption of electricity related to high household income and educational levels; it improved product quality in service industries and enhanced the feeling of security in rural communities. In terms of economic development, the study found electricity important for crop production and food processing industries. The Philippines study was equally positive (Madigan, Herrin, and Mulcahy 1976). For the most part, its findings mirrored those of the Colombia study. It found electricity use associated with higher incomes and education.

Despite the positive findings of the early studies, questions were raised about whether the associations were a cause or an effect of rural electrification. Higher-income households had higher levels of education, and they also had the purchasing power to adopt electricity. Consequently, it was difficult to determine whether rural electrification or income was the main reason for the significant associations. The early studies' methodological drawbacks on the issue of causality left rural electrification wide open to charges by critics who advocated a change in rural energy policy.

Debate and Evaluation within Donor Agencies

At about the time energy was coming under closer scrutiny because of rising oil prices in the late 1970s, criticism of rural electrification became a serious public policy issue. The pendulum began to swing away from rural electrification. The shift began in earnest with two reports critical of rural electrification. In 1977, Development Alternatives completed an evaluation of the program performance of the National Rural Electrification Cooperative Association (NRECA), a trade association group that had planned and implemented many rural electrification projects for USAID. In 1979, Robert Nathan Associates completed a report indicating that the documentation of USAID rural electrification projects was not of sufficient quality to determine whether rural electrification had achieved its anticipated social impacts (Robert Nathan Associates 1979).

McCawley (1979) reviewed the evidence for and against a rural electrification program in Indonesia, concluding that electrification was a "doubtful priority" for rural development and there would be very little chance that the goals of rural electrification projects could be attained. Similarly, Tendler (1979) recommended that the rural electrification policy of emphasizing extension of electricity to rural households should be altered to include a greater focus on rural productivity. Finally, Smith (1980) adopted the position that rural electrification programs were basically uneconomical and unproductive, and renewable energy programs should be adopted instead of grid rural electrification.

Within just a decade, the pendulum had swung from an almost unquestioning acceptance of rural electrification as a necessary development input to a questioning of whether it had any worthwhile socioeconomic impact. While the insights of that period may seem irrelevant today, the extolling of Sustainable Energy for All by the United Nations (UN 2012) and other development agencies suggests that many past lessons may have been lost. This is not an argument against the energy-for-all policies; rather, it suggests that the benefit of hindsight and lessons from the forgotten rural electrification policy debate could inform new policies. The next section focuses on the views for and against rural electrification during the 1980s policy debate.

Advocates

The early justifications of rural electrification were viewed almost as self-evident truths. Rural electrification was considered a necessary condition for future development. At least among the agencies promoting rural electrification, a firm belief prevailed that development could not proceed without electrification. Before the 1980s controversy, in a 1974 report to the World Bank, Gilbert Moon of the NRECA stated the case for rural electrification:

> Once areas have been selected, it appears...that the need for pilot [rural electrification] projects as a basis for experimentation and evaluation is not in keeping with twentieth century expertise. Surely enough is known about the effects of electrification for it to stand on its own merits. Comparative analyses of electrified areas vs. nonelectrified areas may be a boon for statistical researchers, but in the viewpoint of those of us who have spent thirty years in the development of rural electrification, it is a tremendous waste of human resources. Select an area and electrify it on the basis that, if nothing else, it may provide an improved social climate. If the older rural residents make no productive use of electricity, then place your faith in the next generation or the next. (Gilbert Moon and NRECA, *A Report on Rural Electrification Costs, Benefits, Usages, Issues, and Developments in Five Countries* [Washington, DC: IBRD, 1974], 138)

One should not be too critical of a statement made before the 1980s controversy over rural electrification. By 1982, in a talk given to the Society for International Development, a representative of the NRECA had changed stances, stating his firm had benefited from the controversy, which had resulted in studies to evaluate past policies of the organization. Awareness of the controversy led the organization not only to evaluate the efficacy of rural electrification, but also to determine when, where, and with what complementary conditions rural electrification has the most impact.

The following subsections describe the past positions of advocates of rural electrification. Many of the earlier arguments ring true today as international organizations justify policies to support energy for all. It is hoped that the lessons of history will keep in perspective the naive notion that providing electricity for all is a sufficient condition for development.

Electricity as a More Efficient, Less Expensive Option

Efficiency and cost effectiveness are the two most commonly cited justifications for rural electrification. Rural households can benefit from electricity through better-quality lighting and labor-saving small appliances, all at reasonable costs. Small-scale agro-industries can improve productivity through the use of efficient driveshaft power for grain mills and other machines. In agriculture, electric pumpsets can expand both irrigation and food production by enabling farmers to increase yields and grow two or even three crops per year.

Due to economies of scale, central grid power might well be cheaper than available alternatives. In a project appraisal document, the NRECA calculated that, for equivalent household lighting, kerosene for lanterns would cost about twice as much per month as electric lights (NRECA 1976, pp. 33–34). The quality of electric lighting would also be superior to the equivalent light from a kerosene lantern. More recently, this characteristic of electric lighting would spawn new ways to evaluate the benefits of rural electrification (chapter 9). Such labor-saving household devices as electric irons or electric fans would substantially improve the quality of rural life in ways not easily measured in quantitative terms. Nevertheless, the cost competitiveness of electricity versus the alternatives would depend on a host of regional characteristics, including the length of electricity lines and cost of operating them.

Advocates pointed out that, in households that can afford appliances, women and children are the main beneficiaries of rural electrification. Although cooking with electricity is well beyond the means of most rural people, income-producing activities, such as handicrafts, are possible well after dark in households with electricity. Recreational activities like reading and playing games are also possible with good-quality lighting. Several studies found that villagers have a greater feeling of safety when there is public lighting in the small towns and villages (ORG 1977, p. 62; NEA 1978, p. 27). According to advocates, comparisons between light provided by electricity and kerosene favor electricity on the basis of both cost and quality of light.

Electrification's advocates claimed that the use of electric pumpsets for irrigation is more efficient and less expensive per unit of water output than traditional means of irrigation, such as Persian wheels or driving teams of bullocks to haul leather, water-filled bags from a well. Since the timing of irrigation is extremely important, wind power has the drawback of possibly not providing water at the right time. Diesel engines come quite close to the efficiency and (depending on oil prices) cost-effectiveness of electricity, but they are less suitable than electric submersible pumps for deep wells. Diesel fuel is often a foreign exchange drain for many developing countries.

According to the advocates, electricity for driveshaft power and lighting for rural industries is ideal for small rural industries in developing countries. Agricultural processing industries, the main industry in the rural areas of most developing countries, can utilize electricity to process food grains, press oil seeds and perform a variety of other tasks. Beyond food processing, available electrical service may lead to new industries, including small machine shops for repairing implements and cottage industries using electric-powered tools. As in irrigation pumping, diesel engines are an alternative to electric motors, but they require more maintenance, often are more expensive to operate, and may involve foreign exchange problems for purchasing diesel fuel.

Rural electrification advocates stressed that, once a region has electricity service, the number and diversity of rural industries will expand, and the quality of products from existing industries will improve. Rural electrification is a necessary condition for future diversified economic growth. While the impacts may take some time to develop, advocates said that, over the long run, the availability of reliable electrical service will lead to growth of rural industries.

The Urban Bias in Development

The advocates of rural electrification considered it as a way to balance development investments between urban and rural areas. As of 1971, 70 percent of the population in developing countries was rural, only 12 percent of which had access to electricity (table 1.1). For the remaining 30 percent living in urban areas, 80 percent had electricity. Since that time, the situation has changed dramatically. As of 2010, the urban electrification rate was more than 90 percent, while the rate in rural areas had surpassed 60 percent (table 9.1). Despite this progress, about 1.3 billion people remain without electricity. Because today's households without electricity are poorer and more remotely located than those of 40 years ago, the challenges and controversy over rural electrification are even more difficult now than in the past.

Despite this large disparity for 1971, the World Bank estimated at that time that only about 10 percent of the cumulative investment in the power sector was for rural electrification (World Bank 1975, p. 17). Electrification's advocates pointed out that this bias was compounded by food-pricing policies in many developing countries that held down the cost of food for urban populations. Thus, the rural areas provided less expensive, and some might say subsidized, food to urban areas. In addition, a large percentage of development programs were for capital-intensive projects in urban areas. For instance, lending for rural development was increasing substantially at the World Bank in the 1970s, but only 20 percent of total loans for the 1968–74 period were for agriculture or rural development (World Bank 1975, pp. 86–87). Rural electrification would be a convenient way to redress some of the urban-rural inequities in development programs. Since it cost less to provide service to high-density urban regions than to sparsely populated, rural areas, revenues from the more profitable urban electrification could subsidize rural electrification in developing countries and thus balance food-pricing policies.

Table 1.1 Extent of Rural Electrification in Developing Countries by Region, 1971

| | Population | | | Electricity access | |
| | | | | Rural | Rural |
Region	Total (million)	Rural (million)	Rural (%)	Rural population (million)	Population (%)
Latin America	282	140	50	32	23
Asia	934	700	75	105	15
Africa	182	165	91	7	4
Other	143	87	61	45	15
Total	1,541	1,092	71	189	12

Source: World Bank 1975.
Note: An update of these historical figures is provided in table 9.1.

On the question of equity, advocates asserted that rural electrification could directly and indirectly benefit the rural poor and redress regional inequities. The rural poor who adopt electricity service would experience improvements in their quality of life. Compared to wealthier households, the rural poor would be less likely to adopt electricity; however, low-income families often sacrifice a substantial percentage of their earnings to obtain the benefits of electrification. A 1980 survey reported that 15 percent of the poorest rural households in the Philippines had electricity, a surprising number given the low levels of income available for basic needs at the time (U.S. Bureau of the Census 1981, p. 18).

Advocates also emphasized that electrification indirectly increased employment and spurred various income-producing activities. For instance, with electricity, land could be farmed several times a year, substantially increasing the amount of required labor. Capital improvements that increased productivity created additional demand for labor. While agriculture was still the main source of economic activity in rural areas, new small-scale industries using electricity created jobs. Thus, while most of the rural poor probably did not benefit directly from rural electrification, according to advocates, employment of the rural poor in industry and agriculture-related jobs was a significant indirect benefit of electricity.

In short, advocates of rural electrification claimed that such investments were infrastructure improvements that would enhance the quality of rural life and lead to long-term increases in rural productivity and employment. The result would be that rural areas would become more desirable places to live. Improvement in the quality of rural life could have innumerable spillover effects, such as reduction in the extent of rural-urban migration and greater willingness of qualified professionals, including doctors and agricultural extension workers, to serve in rural areas.

Critics

The belief that rural electrification was an appropriate development strategy for almost all developing countries was shaken during the early 1980s. Critics asserted that it was too expensive, did not benefit all social classes equally, and had no direct impact on agricultural or industrial development. This all-encompassing criticism brought into question the funding for rural electrification.

During the 1980s, funding trends for rural electrification involving large projects were hard to discern (Niblock 1982). Over the previous two decades (1965–85), USAID had been the dominant funding agency for rural electrification projects. In the early 1980s, however, it completed just two projects: one in Bangladesh and another in Pakistan. Throughout the 1970s, the World Bank had steadily funded rural electrification, and during the 1980s, there was no curtailment of its lending activity. The Inter-American Development Bank (IDB) was the third major donor agency involved in funding rural electrification projects. IDB lending increased by over US$100 million for the 1965–70 to 1970–78 periods, but then declined by about $100 million during the 1975–79 period (IDB 1979b). This change in lending figures perhaps was due to the ebb and flow of new projects. For most other donor agencies, funding for rural electrification was sporadic, so no over-time trends were discernible.

The basic criticism of rural electrification was more philosophical in nature, and persists even today. It centered on the place of rural electrification in energy strategies for socioeconomic development. The questions raised concerning rural electrification centered on the socioeconomic impact and cost effectiveness of investments. The five central issues raised by critics were as follows:

1. The appropriate time for implementing rural electrification projects in the development cycle;
2. Whether rural electrification programs should have a household or rural productivity emphasis;
3. Whether the rural poor benefit;
4. The capital-intensive nature of investments; and
5. The place of rural electrification in more general energy strategies for economic development.

Rural electrification by itself obviously will not irrigate fields, apply fertilizer or produce industrial goods. Critics pointed out that rural electrification had to be placed in the context of complementary rural development programs to have a substantial impact in the countryside.

Is It Too Soon?

The success of rural electrification in the United States during the 1930s provided the model for extending electrification to developing countries. The critics questioned whether this experience was relevant to poor rural areas in developing countries. The argument was that the growth of household adoption in the U.S. probably could not be replicated in developing countries because of the substantially lower levels of household incomes and rural employment. As evidence, opponents pointed out that projections of load demand by utilities in some countries were disappointing (Sen Gupta 1977; Sambrani et al. 1974a, b).

After a thorough review of projects in Indonesia, McCawley (1979, p. 68) criticized rural electrification as doing little to achieve expected benefits. He reasoned that electricity prices were generally set so high that only the wealthy could afford electricity. The fact that low incomes constrained participation by the poor in rural electrification programs caused a cascading effect of lower than anticipated growth in connections, putting undue financial strain on the utilities. McCawley concluded that other developmental priorities should take precedence over rural electrification.

Table 1.2 Access to Electricity in Developing Nations, 1980

Country	GNP, 1978 Dollars per capita	People with access to electricity (%) Rural	Urban	Total
Suriname	2,110	44	88	61
Argentina	1,910	5	99	80
Uruguay	1,610	65	95	80
Brazil	1,570	5	95	62
Costa Rica	1,540	52	95	70
Chile	1,410	30	99	85
Mexico	1,290	50	99	80
Korea	1,160	90	99	95
Jamaica	1,110	15	60	35
Ecuador	910	30	72	38
Colombia	870	19	86	62
Ivory Coast	840	20	90	45
Nicaragua	840	20	64	41
El Salvador	600	12	75	36
Bolivia	510	10	78	33
Philippines	510	--	--	31
Thailand	490	18	70	27
Honduras	480	--	--	21
Cameroon	460	0	28	6
Liberia	460	2	85	25
Kenya	320	--	--	6
Pakistan	230	--	--	22
Tanzania	230	--	--	7
Sri Lanka	190	1	23	6
India	180	--	--	14
Chad	140	6	70	10
Ethiopia	120	0	33	4
Nepal	120	--	--	5
Bangladesh	90	0	19	2

Source: World Bank 1984.

Note: (--) indicates data not available.

Just as today, in 1980 countries with relatively high per capita income had higher rates of rural electrification. Along with electricity percentages, countries were arranged by gross national product (GNP), a general measure of income (table 1.2).

Access to electricity services was generally lower in such poor countries as Burma, Chad and Ethiopia, while the reverse was true in higher-income developing countries, including Costa Rica, Chile and Mexico. In the 1980s, few African countries had extensive rural electrification programs, a pattern that continues even today.

The percentage of rural populations that adopted electricity when it became available was also higher in wealthier countries (table 1.3). In a wealthy region of Argentina, for example, a rural electrification cooperative was able to connect more than 90 percent of rural households, even though some were widely scattered on farms and ranches. The mean per capita income for Argentina in 1978 was US$1,910 per person, well above the average for most developing countries.

Table 1.3 Gross National Product and Household Electricity Adoption in Areas with Electricity, 1978

Country	GNP, 1978 (dollars per capita)	Households adopting electricity in areas with electricity (%)
Argentina	1,910	over 90
Costa Rica	1,540	70-80
Ecuador	910	35-64
Colombia	870	50-68
Philippines	510	42
Philippines	510	20-51
Bolivia	510	12-22
India	180	26

Sources: Kessler et al. 1981; Goddard et al. 1981; Butler, Poe, and Tendler 1980; Mandel et al. 1981; NCAER 1981; U.S. Bureau of the Census 1981; Argentina, personal communication with IDB; World Bank 1980.
Note: Two Philippines values are for two studies.

In the Philippines, which had a US$510 per capita income, the adoption rate for customers was about 42 percent of potential customers (U.S. Bureau of the Census 1981). A survey in Colombia found that only 12 percent of potential customers in the lowest income group adopted electricity service, compared to 68 percent for all households surveyed (Saunders et al. 1978). The mean per capita GNP for Colombia in 1978 was $870, above average for developing countries. Such evidence spurred critics to suggest that, for most developing countries, it is too soon for rural electrification, and other development programs should be commenced prior to rural electrification in order to raise income levels. From this viewpoint, rural electrification is a social luxury that should be implemented at a later stage of development.

Residential Versus Productive Use of Electricity

In the 1970s, many countries emphasized residential electrification, which became a point of sharp contention. Critics argued that, since wealthier households could easily afford to adopt electricity, rural electrification programs with a household emphasis were really a subsidy for the wealthy in rural areas.

The critics contended that the rural poor had not benefited from programs that provided electricity to wealthier households (Tendler 1979, p. 21). Instead, gains from electrification for the poorest could occur through productive uses of electricity to increase employment. In addition, public uses of electricity for health centers and village lighting would have greater benefits for the poor than electricity programs that stress household electricity use. Tendler concluded that the focus of rural electrification should not be on subsidizing the rich, but stimulating economic development and rural employment. If rural electrification could cause significant job growth, then perhaps the rural poor could benefit from electricity programs.

But can a country demonstrably alter the impacts of rural electrification by changing the focus of the program from household to rural productivity? The answer really depends on many conditions, including the level of a country's development and the extent of commitment to allocate funds to improve rural productivity. The two contrasting cases of India and the Philippines in the 1980s are instructive. As previously indicated, in the Philippines, rural electric cooperative have promoted residential household electrification. In 1975, 10 rural electric cooperatives sold about 48 percent of their kilowatt-hours to residential customers (Denton 1979, pp. 138–43) (table 1.4). Only 4 percent of sales were for agriculture, and industrial sales were approximately 10–18 percent. The 1982 figures for rural electric cooperatives in the Philippines indicate the situation had changed. There had been a rise in the share of kilowatt-hour consumption for industry to 27 percent and a decline for households to 37 percent, but agriculture's share had remained quite low.

The NRECA, which has developed, designed and promoted numerous rural electrification projects in developing countries, had recommended a household emphasis and administration of programs by rural electrification cooperatives projects. The NRECA philosophy then and even today is to support extensive coverage of both rich and poor rural households and administer programs through consumer-oriented rural electric cooperatives. This emphasis led to high household connection rates for countries like the Philippines and Costa Rica. Thus, the rural electrification schemes designed to promote household coverage in most cases reached their goal (table 1.4).

The household emphasis in the Philippines contrasted sharply with the situation in India in the 1980s. During those times and even today, agricultural connections are heavily subsidized and residential users must pay higher rates. In villages with electricity in India during those times, very few rural households had electricity. Irrigation pumping accounted for over four-fifths of the total electricity load (Sen 1980, p. 28). Of the remaining one-fifth, just over one-tenth went for domestic lighting and the rest was for rural industry and commerce. Similarly, over 50 percent of electricity use went to agriculture and less than 20 percent was for household consumption (table 1.4). At least for India and the Philippines in the 1980s, it appears that during the rollout of programs, a household or productivity policy emphasis had an impact on whichever sector had the greatest connected load, although over time these differences eventually diminished.

Table 1.4 Type of Electricity Consumed in Rural Areas, 1970s

Country	Electricity consumed by sector (%)				
	House-hold	Com-merce	Indus-trial	Irriga-tion	Other
Colombia (1971)	55	27	3	--	15
Costa Rica (1973)	28	--	--	--	72
El Salvador (1972)	36	24	34	2	4
Nicaragua (1976)					
COERAM	30	-	2	60	8
CODERSE	26	4	55	4	11
CAEER	54	6	18	2	20
India					
Telanga PP (1975-76)	21	12	17	48	2
Suryapet, A.P. (1971)	6	1	4	88	1
Una, Gujarat (1973)	13	--	7	79	1
Bayad·Modasa, Gujarat (1973)	18	--	23	54	5
Indonesia (1974)	69	--	3	-	28
Philippines					
Camarines Sur I (1975)	60	29	--	1	10
Albay Coop. (1975)	64	21	4	--	11
Misamia Oriental (1975)	44	16	23	2	15
Thailand. PEA (1972)	30	34	34	--	2

Source: Adopted from Cecelski with Glatt 1982, p. 10.
Note: All rows are equal to 100 percent; (--) indicates either 0 or not available.

The critics also argued that the benefits of electrification for agriculture and small-scale industries also go largely to individuals in upper-income classes. Small farmers must be able to afford the capital investment in pumpsets or other complementary inputs. Rural artisans need to be able to purchase the electrical equipment that will improve their productivity. The expected result was that people in the upper strata of rural communities would be able to take advantage of the benefits of subsidized electricity. By contrast, the poor would continue to rely on traditional and noncommercial sources of energy (Sen 1981).

Community and Regional Inequality

Rural electrification might also cause regional imbalances in socioeconomic development. Not only does it disproportionately benefit the wealthier rural households; critics also contended that rural electrification would primarily benefit a country's wealthiest regions. The reasoning was that capital and complementary inputs are necessary before rural electrification can have a significant impact on socioeconomic development, and such inputs have a greater impact on productivity and employment generation in more prosperous regions. A 1980s study in the Philippines concluded that "the contribution of electricity to the development process depends on the level of development of the area, the availability of capital and other financial and human resources as well as the implementation of programs which stimulate the use of power" (Mandel et al. 1981, pp. 13–14). Another 1980s study of a backward region in India found that the productive use of electricity lagged behind expectations owing to a lack of capital, inability to obtain bank loans, insecurity of markets for products and unreliable supplies of electricity (Center for Studies in Decentralized Industries 1980). This led critics to assert that electrification had only limited social benefits in poorer areas, but could be used in more productive ways in prosperous regions of developing countries, thus exacerbating regional inequalities.

Too Expensive and Little Productive Impact

Rural electrification programs are quite capital intensive. Construction of generating capacity and extension of lines into rural areas are usually subsidized by governments in developing countries. Concessional loans from donor agencies and multilateral banks have been necessary to make the financing of rural electrification attractive to electricity companies. Such large investments can be justified only from the long-term stream of revenues and the social and economic benefits produced by projects.

Even today critics argue that grid rural electrification projects have little impact on rural productivity in agriculture or small-scale industry or on regional levels of socioeconomic development. In an article advocating decentralized energy for rural communities, Smith (1980) asserted that rural electrification had no productive benefits and generated no additional regional development. Smith observed that most industries that were started after regions or villages were provided electricity may have done so for independent reasons; moreover, they could have been powered by another energy source. The conclusion of the critics' line of reasoning was that the economic benefits of rural electrification for establishing rural industries were insignificant when compared to the costs involved in extending the central grid. The same arguments were made about the effects of rural electrification for irrigation. Critics said that individual diesel engine pumpsets could perform the same function as electric motors without the substantial investments necessary for constructing generating capacity and expensive distribution systems.

During the 1980s, critics asserted that investments in rural electrification were investments forgone for other community services or energy programs (Lovins 1977). For instance, compared to rural electrification, investments in schools, water supply, health services and roads all could have substantially greater direct impacts on literacy, population growth, rural-urban migration and other social benefits. Critics believed the opportunity costs of rural electrification far outweighed the small, insubstantial social benefits. They argued that money spent on other rural development or alternative energy programs could more equitably and effectively improve rural productivity and quality of life.

Alternative Energy Strategies

In the 1980s, critics often blamed rural electrification for preventing a more balanced energy development strategy that would include traditional fuels (e.g., fuelwood and charcoal), along with alternative technologies (e.g., biogas and wind energy). Advocates of rural electrification were accused of recommending their technology for anyone and every location, regardless of the availability of energy resources in villages (Development Alternatives 1977). Before the energy crisis of 1973–74, electrification was the only significant rural energy development strategy. By 1980, rural energy needs were being examined more critically, which made expensive grid rural electrification programs easy targets for criticism.

Many of these arguments against grid electrification ring true today, particularly with the international and donor emphasis on sustainable energy for all (UN 2012). The argument over whether rural people should have electricity is over, but the issue of how to accomplish electrification is ongoing. The current emphasis on sustainable energy is not new; it was discussed as early as the 1970s, with Lovins' classic study, *Soft Energy Paths* (Lovins 1977).

Decentralized or "soft paths" of energy development typically involved matching the energy end use with the energy source (Lovins 1977; Pendse 1980a, b; Smith 1980). During those times, as today, many critics of grid rural electrification were in favor of alternative technologies suitable for specific locations and end uses, including wood or biogas for cooking or hydropower for irrigation. The type of technology would depend on the resources available in a village. Obviously, a village located on a semi-arid plain could not adopt hydropower as a viable energy source. Alternatives to central grid electricity were still in experimental stages, but they raised issues that are still debated today. The question was this: At what stage of development or level of population density do alternative technologies make more sense than central grid rural electrification.

Conclusion

B y definition, the main goal of all development strategy is social and economic development. The rural electrification controversy of the 1980s revolved around some of the same issues facing energy development planners today, including choosing a strategy to reach the goal of providing modern energy for all. The advocates of rural electrification strongly believed that rural electrification was a necessary condition for development and that it would be a catalyst leading to improved rural productivity and a better rural quality of life. Its critics argued that other projects should receive priority over grid rural electrification in developing countries, especially in those countries at a lower level of economic development.

The arguments against rural electrification were that the current scale of projects was too large; the extension of the grid to isolated rural areas too costly given the weak demand; and the predominant focus in some countries on providing electricity to rural households rather than productive activities inappropriate. But from a policy standpoint, rural electrification certainly can complement other technologies.

Many of these 1980s arguments against rural electrification depended on uncertain factors. One factor was the volatile price of oil. Even today, the uncertain price of petroleum fuels complicates the picture for both central grid and decentralized alternatives for rural areas. Another factor was that alternative energy schemes, such as biogas, wind and direct solar alternatives, must be practical and cost effective. Today, defining how these technologies fit into energy-for-all strategies remains a challenge. Finally, there was the question of the actual socioeconomic impact of rural electrification. In order to sort out the claims and counterclaims by advocates and critics, this study put rhetoric aside to examine the impact of rural electrification in the development process.

2. Impact Evaluation Study Design

Whether rural electrification is an appropriate strategy or even a catalyst for rural growth has been a matter of public policy debate for many years. Even today many international organizations advocate for rural electrification, but fail to weigh in on how to improve the impact of programs for the over 1 billion people still without electricity. The past debate between advocates and critics of rural electrification centered on the best means to achieve socioeconomic development. Without question, electricity is a future necessity for rural areas in developing countries. The debate centers on whether the emphasis of programs should be residential or productive or whether electricity introduces greater inequalities into rural areas since typically the wealthiest households are the first to receive a connection.

Some fundamental issues being debated involve disagreements over the socioeconomic outcomes of rural electrification and whether the high capital investments in electricity distribution are justified. One might imagine that many differences of opinion could easily be resolved by simply quantifying the benefits, assigning them monetary value and weighing these against the cost. Unfortunately, resolving the outstanding issues of the debate is not so simple. Many of the socioeconomic consequences of rural electrification are difficult to quantify, especially in monetary terms. For example, a higher quality of lighting or changes in quality of life are hard-to-measure benefits. Also, assessing the costs of rural electrification is complicated by electricity growth rates, load factors and many other considerations that vary by country.

Early on in the research it was recognized that rural electrification varied substantially between countries. To avoid the pitfalls of assuming that country-specific characteristics are general consequences of rural electrification, a comparative research design was adopted to address issues involved in the impact of rural electrification. To this end, detailed rural electrification evaluations were completed in the 1980s in three countries: India, Indonesia and Colombia. These countries have distinct histories of rural electrification and industrial structures. For example, the focus of India's rural electrification program originally had been to improve agricultural productivity, while the emphasis in Colombia and Indonesia had been on rural households. Although the most comprehensive data from the empirical studies are from India, the country comparisons provide interesting contrasts on the impact of public policy on the outcome of rural electrification programs.

Rural electrification projects affect countries at all levels: regional, community and household. A unique feature of this research is the use of three units of analysis—rural industries, communities and households—which allows for the examination of macro-level changes. Most previous studies had focused exclusively on households, which can be inappropriate for examining changes in literacy (individual child), migration (village) or inter-regional equity (regional differences). The expansion of empirical knowledge concerning the impact of rural electrification on macro-changes in the structure of rural villages is one of the important contributions of this research.

An examination of all the policy issues surrounding rural energy and rural electrification is beyond the scope of this book. Nevertheless, the question of rural electrification's impact on long-term socioeconomic development, equity, and rural quality of life is analyzed thoroughly, as are the actual costs and subsidies involved. The next sections describe the questions deemed most important in the rural electrification debate and the regions where the three surveys were conducted. These are followed by a detailed description of the organization of the research.

What Is Rural Electrification?

Defining rural electrification would appear to be an easy task. Obviously, there is rural electrification when rural households or communities have electricity for use in homes and businesses and on farms. However, electricity may be distributed unevenly. For example, if four households out of a community of thirty households have electricity, does this mean the community has rural electrification? Is a village classified as having electricity if it has service at night, but not during the day, as is often the case for communities with isolated diesel-powered generation? Another dimension of the issue is that electricity can be generated from a variety of sources. A large central grid may use coal, oil or hydropower, while small decentralized sources typically include mini- and micro-hydropower plants, diesel plants, biogas generation units, photovoltaic cells and wind energy.

Most rural electrification programs in developing countries involve some form of connection with a central grid. For densely populated countries with emerging grid networks, this may be the best way to plan rural electrification schemes. However, there are some important exceptions. In China, for example, micro-hydro supplies a significant portion of the electricity for rural areas, although these electricity generation sources today now are connected to the central grid. In the South Pacific, some islands are supplied almost entirely by local generation sources; because they are so widely scattered, any grid-based system would be virtually impossible. The issues of electricity generation and the most efficient means of distribution are extremely important for determining electricity cost, a subject addressed in chapter 7.

For assessing the impact of rural electrification, the generation source does not matter unless cost or technical considerations place constraints on the extent of service and the time that electricity can be consumed. The benefits of household lighting undoubtedly will be the same no matter whether the electricity is generated at a coal-fired plant 100 kilometers from a house or from a photovoltaic cell placed on the rooftop. However, the extent to which electricity is used in a community is extremely important for evaluating its impact, and this may be limited by the technology used to produce electricity. A village in which four households have had electricity for one year would differ from a village of similar size where fifty households have had electricity for ten years. Although electricity is basically the same no matter what the generation source, its present and future development can be affected by limitations on service.

In this book, rural electrification is defined as the availability of electricity for use in rural communities, regardless of the form of generation. The impact of rural electrification can vary according to the number of years that a community or household has had electricity, the extent to which it has been adopted in the community for various activities and the quality and extent of service (chapter 7). The source of generation does involve issues of cost and quality of service; but given the same extent and quality of service, the source of generation has few consequences for the impact of electricity on rural people. For most of the communities and households surveyed in this book, rural electrification means the extension of the central grid to rural areas. Nevertheless, similar impacts could be achieved if similar levels of electricity were generated from isolated, decentralized sources.

Research Questions

A general set of questions and objectives were formulated at the inception of the research to guide and coordinate the empirical investigations. These research objectives were developed after a thorough review of the literature (Cecelski with Glatt 1982), as well as extensive interviews with donor agencies instrumental in funding rural electrification. The questions and issues listed below are part of the study's original research objectives, and, to some extent, are reflected in the chapter organization of this book; they are as follows:

1. Does rural electrification generate additional productivity, higher income, greater employment and structural change in rural areas? Does it contribute to greater regional equity and reduce excessive migration to urban areas?

2. How does rural electrification fit into a broader strategy of rural development? What complementary conditions or inputs make for success or failure? Can one rank the complementary conditions and inputs in order of importance and evaluate their relative importance to rural electrification itself?

3. What are the effects of rural electrification on "equity" in development (e.g., providing benefits to different income classes or widening opportunities for small farmers, landless agricultural workers and artisans)? How does rural electrification affect the roles of women and children?

4. What are the benefit-cost ratios of rural electrification in (a) financial terms and (b) economic terms?

5. What are the most effective types of rural electrification policies and strategies to maximize the overall benefit-cost ratios (including priorities in selection of project areas, subsidies and tariff structures)?

6. How does central-grid electrical service for rural areas compare to decentralized forms of electricity/energy production, including biogas, wind and other forms of energy?

7. How do different rural electrification policies affect the impact and implementation of rural electrification in various countries?

Needless to say, not all of the above questions can be answered definitively. Yet most have been explicitly analyzed using either quantitative analysis or more qualitative methods. The unifying themes behind these research objectives are economic development, rural quality of life and the costs of rural electrification.

Description of the Sample Areas

The study of the impact of rural electrification is based on surveys conducted in the 1980s in India, Colombia and Indonesia. Where appropriate, the findings have been updated with results from more recent studies. Within these countries, communities and households were selected based on their level of development and number of years with electricity. For both India and Colombia, the states or regions in the sample were quite diverse, while the sample for Indonesia was from one relatively homogeneous region. In India, the four sample states were located in geographically and culturally distinct regions, and the same was true for Colombia. The regions selected in India ranged from the highly developed, groundwater-irrigated Punjab to the dry-farming, pulse-growing plains of Maharashtra. In Colombia, the mountain coffee-growing regions contrasted sharply with the traditional small-farmer central areas. The following paragraphs generally describe the salient features of the study regions. (The Appendix provides a detailed description of the research design for the countries).

The sample areas in India were located in the four major regions of the country (figure 2.1). The surveys were finished in 1981.[1] Andhra Pradesh, a state in South India, is bordered on the east by the Bay of Bengal. Many parts of the state are heavily canal-irrigated and

---

[1] The India, Colombia and Indonesia surveys are henceforth referred to as follows: India Survey 1981 (see Samanta and Sundaram 1983), Colombia Survey 1981 (see Velez with Becerra and Carrasquilla 1983) and Indonesia Survey 1981 (see Brodman 1982).

are productive rice-growing regions, yielding two and, in some cases, three crops or harvests a year.

Figure 2.1 India Survey Area, 1981

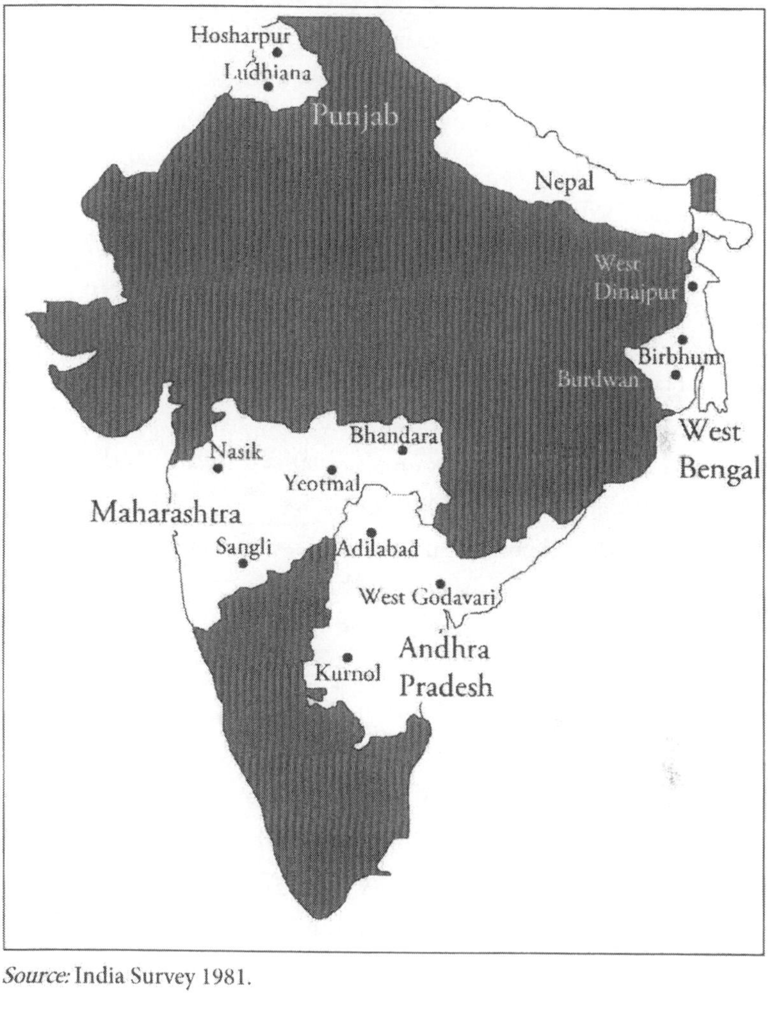

Source: India Survey 1981.

While primary crops are grown in Andhra Pradesh during the monsoon season, a second crop is grown on the moisture left after the first harvest in some of the irrigated regions. In contrast to the productive rice-growing areas, the poor drier regions grow coarse grains and pulses, requiring irrigation for growing more than one crop per year.

Maharashtra, a neighboring state in West India, extends all the way to the Arabian Sea. Except along the coast, the state has somewhat less rainfall than Andhra Pradesh, but canal irrigation is not as well developed. Most farmers rely on small pumpsets powered by either diesel engines or electric motors linked to an irrigation pump. As a result, crops are grown during the monsoon and again during the winter season when no rain falls. Maharashtra has a somewhat lower agricultural income per person than Andhra Pradesh.

The other two states in the India survey, Punjab and West Bengal, both feature high agricultural yields. But for the years before the survey in the 1980s, agriculture had advanced tremendously in the Punjab, while it had stagnated in West Bengal. In fact, these two states are sometimes compared to draw lessons for agricultural development, with Punjab cited as the model to follow.

Punjab is agriculturally the most advanced state in India and now exports both rice and wheat to other states. Land in Punjab is irrigated from private tubewells, and crops are grown during two and sometimes three seasons. Although the major crop is wheat, production of winter rice has been expanding significantly. By contrast, West Bengal, historically one of the most productive states in India, has not kept pace with the nation in terms of crop yields. Although the state features rich soils and high annual rainfall, it has few private wells, so double-cropping is quite low. Paddy rice grown during the monsoon season is its staple crop. The high yields per acre in West Bengal are offset by its high population density; the agricultural income of US$63 per person compares unfavorably with $144 for Punjab. Thus, the four states in the India survey are clearly quite diverse and represent regional differences in that country.

In Colombia, three diverse regions were selected (Velez with Becerra and Carrasquilla 1983): the North Coast, typified by large cattle farms; the Central Region, with small traditional farms; and the coffee-growing area (figure 2.2). The North Coast region is sparsely populated and has large and, in many instances, modern farms. However, the region features extremes in socioeconomic levels of living.

Figure 2.2 Colombia Survey Area, 1981

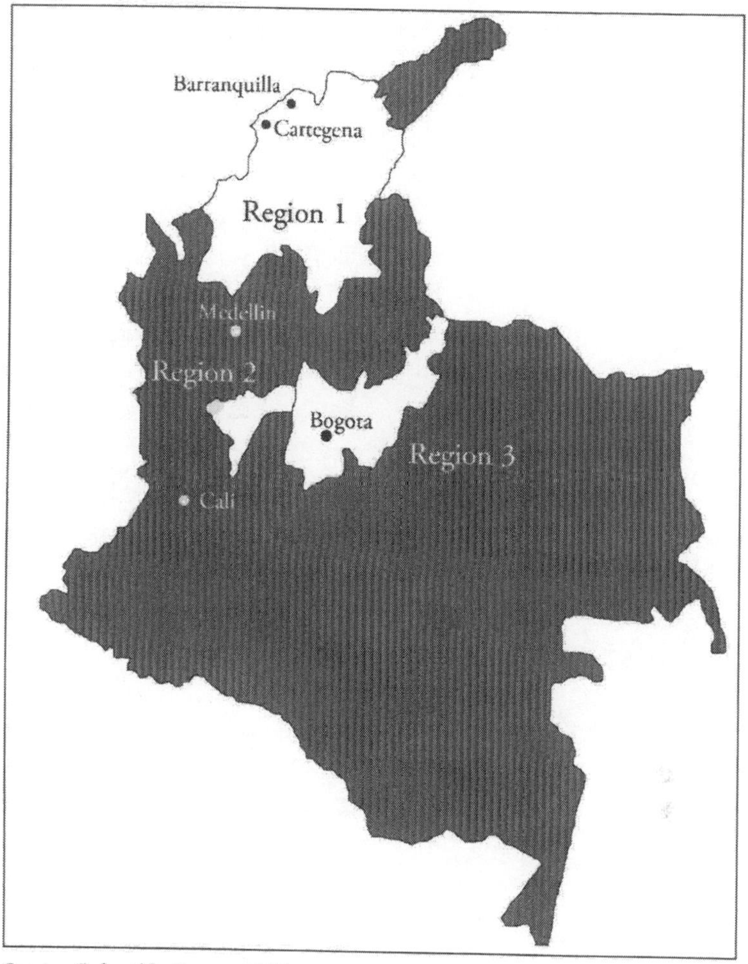

Source: Colombia Survey 1981.

The coffee-growing region, characterized by both large and small plantations, is more densely populated. Since coffee is Colombia's main export crop and a major source of foreign exchange, the coffee region has the country's highest standard of living. One of the region's unique features is the significant investment made by the National Federation of Coffee Growers in education, health and other "basic needs" programs. Thus, the area ranks high for most social indicators. Although the Central Region contains the city of Bogota, the rural areas are quite mountainous, with traditional agriculture dominated by small farms. Compared to the North Coast, the coffee region has more evenly distributed income and a standard of living comparable to the national average.

The Indonesia sample area differed from those of India and Colombia. The selected district of Klaten became a pilot area for a rural electrification project in 1977 (figure 2.3). At the time of the survey, Klaten was a relatively advanced region. Its productive agricultural base was dominated by rice produced on land endowed with year-round, gravity-fed irrigation. Somewhat atypical of Indonesia, Klaten had a vigorous household enterprise sector, along with small-scale industries and businesses.

Figure 2.3 Klaten Survey Area in Indonesia, 1981

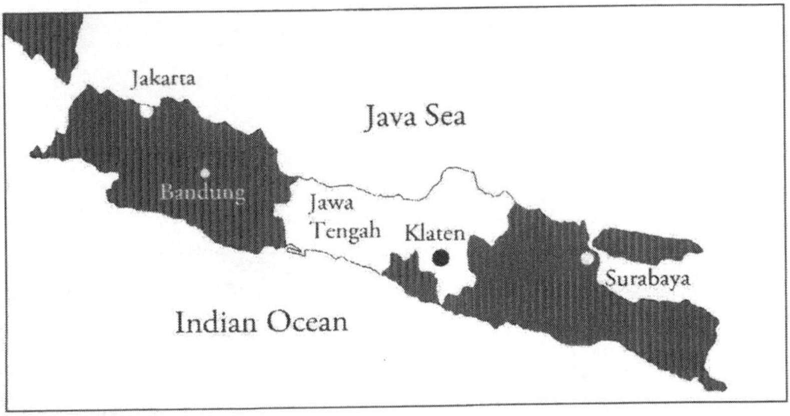

Source: Brodman 1982.

Klaten is located only 60 kilometers from the district capital, a moderate-sized rural town serving as a marketing center for the surrounding agricultural community. Like the rest of the district, the eight villages in the study were situated in one of the world's most densely populated agricultural areas. Compared to the national average, they were wealthier and had more industry. Thus, the findings from the Indonesia study were qualified by the fact that the region was highly developed and had only recently received electricity.

Organization of the Research

T he rural electrification public policy debate during the 1980s presented a dilemma for developing countries. There was growing evidence that the beneficial impact of rural electrification in the United States would not be duplicated in developing countries, especially in those with low per capita income. However, there were still many unresolved questions. Which of the critics' charges were mere accusations? Should rural electrification projects be dropped completely, cut back or altered to have greater socioeconomic impact?

The resolution of the 1980s controversy depended on an evaluation of the socioeconomic benefits, along with the cost of rural electrification as examined in this research. During the 1980s the productive consequences of rural electrification needed to be evaluated over the long term. However, at that time many evaluation studies of rural electrification were limited to examining villages or communities only a few years after they received electricity service, thus measuring only short-term effects.

The productive use of electricity had to provide a principal justification for rural electrification, even if there were major social benefits. The major contributions that rural electrification could make to rural economies were gains in productivity and income and the creation of jobs, both for agriculture and industry. Even when only a minority of rural households connected to the electric grid, the projects could be worthwhile if there were jobs gains and higher incomes for the rural poor. While electrification of households might have directly benefited only village elites, it also may have helped to expand irrigation, which ultimately stimulated employment and agricultural productivity. This 1980s research examined in what ways and context rural electrification had the greatest impact on rural productivity, incomes and employment. Also discussed was whether the changes in rural productivity, if any, justified the high costs and large subsidies incurred by rural electrification.

Agriculture and Industry

In most developing societies, agriculture employs as much as 80 percent of the rural population. The primary benefit of rural electrification for agriculture is the expansion of irrigation capacity. Rural electrification may indirectly stimulate increased rural employment by expanding cultivated land through double and triple cropping and working land already under cultivation more intensively. Areas that have participated in the so-called Green Revolution also have reported increases in rural employment, at least in the initial stages. Electric-powered machines substitute for labor in bottleneck periods, such as harvest time. Small-scale threshers might facilitate harvesting so that new crops can be planted for a second growing season. One caveat is that diesel engines may also have the same benefits attributed to rural electrification for irrigation.

The impact of electrification on rural industries may be less pronounced than for the agricultural sector. In the past, agro-industries often relied on diesel engines or animate power. The most noticeable consequence of rural electrification may be the substitution of electric motors for such diesel engines. Even though electric lights might increase their productivity by allowing work to continue after dark, many crafts and household industries currently do not absolutely need electricity directly for production. Most medium-size and major industries have already chosen to locate in cities, towns or growth centers because of the availability of electricity, transportation and communication facilities.

Social Effects

Rural electrification can have structural impacts on rural communities, as well as intra-household effects. Many studies have exclusively examined rural households for only the direct impacts of rural electrification. However, rural-urban migration, population growth, education and literacy are all structural characteristics of communities that can be affected. Even today, little research has focused on the relationship between rural electrification and migration. Public meeting places, warehousing, night schools, and street lighting are community services that might be initiated or improved with rural electrification. Although these are not direct productive consequences, they nevertheless may have a considerable effect on community life and indirectly may impact rural productivity. Likewise, rural electrification can adversely affect rural communities.

At the household level, electricity can change rural living and energy-use patterns, but electricity benefits may accrue only to wealthier households. For instance, rural electrification can provide clean, high-quality lighting in homes. When households first switch to electricity, they discontinue the use of kerosene lamps or lanterns in favor of electric lights. The purchase of other household items (e.g., fans, black-and-white television sets and small appliances) may follow lighting, thus altering intra-household work and living patterns. But if electricity is only used in high-income households, then program benefits may go only to a community's wealthiest residents.

Benefits, Costs and Rate Structure

Cultural factors and traditional practices are apparently not a major constraint for rural electrification. The major stumbling block to future growth is the financial viability of electricity companies. The limited use of electricity in rural areas and thinly spread consumer demand mean that developing country governments often subsidize rural electrification. Political resistance may develop against indefinite subsidies to the electric utilities, and, in some countries, the question of whether rural electrification can be self-supporting now or in the distant future is still unknown. One major problem thus concerns the subsidy policies and tariff levels to keep the industry viable. The organizational, rate structure and subsidy questions relate essentially to the electricity industry, but have broad ramifications for development strategy.

The costs of alternative power-generation schemes need to be compared with those of central grid generation. Tendler (1979, pp. 44–49) made a strong case for diesel generation systems for residential service. The advantage of local generation is the manageability of maintenance problems. Another benefit is that capital investment can be spread over a longer period of time. Furthermore, systems are not required to run 24-hours a day, which reduces costs substantially. But even Tendler recognized that, in the long run, diesel generation was more expensive than central grid electricity. Other potential decentralized sources of power generation include diesel engines adapted to run on biogas fuel, wind power, photovoltaics, mini-hydro, micro-hydro and other alternative sources. Some newly emerging technologies have very small lighting systems and a reasonable cost. Thus, it is important to evaluate the comparative costs of central grid versus decentralized generation for providing electricity service to communities in rural areas, especially those in more remote regions.

Conclusion

The rural electrification debate was the beginning of new views on developing rural energy strategies for developing nations. The consequence of the debate was that staunch advocates reduced their claims regarding the benefits rural electrification for development. However, they continued to acknowledge rural electrification as a necessary condition for social gains and economic growth. The severest critics shunned rural electrification as ineffective for development; even today, many advocate an energy policy based on new and renewable technologies. Between these extreme positions, opinions have varied widely, ranging from an emphasis on rural productivity to decentralized generation of electricity and postponing rural electrification programs in nations with extremely low levels of economic well being. These arguments are still relevant today as development agencies recommend ways to provide energy for all.

During the 1980s, the paucity of rigorous empirical studies on rural electrification meant that arguments for and against it were marked by a lack of consensus on key socioeconomic issues. Many of the complex, and sometimes controversial, questions surrounding rural electrification remain quite relevant today. Some of the items debated involved ideological beliefs about the nature of development, while others reflected the complex nature of the development process. But even on topics for which information has been less than ideal, judgments have been made, taking into consideration the most recent studies on the impact of rural electrification on development. New insights were gained from the India, Colombia and Indonesia studies. Those comparisons, along with more recent research, offer insights have would not have been possible from examining the countries individually.

3. Agricultural Development in India

The 1980s controversy over the role of rural electrification in development was in part due to conflicting impact evaluations. In some developing countries, electricity apparently had little impact on agriculture (Saunders et al. 1978; Butler, Poe, and Tendler 1980; Mandel et al. 1981; Goddard et al. 1981; Madigan, Herrin, and Mulcahy 1976), while in other countries the impact was substantial (Samanta and Varma 1980). The disagreeing evaluations generated a clear need for more research on whether rural electrification stimulated agricultural development, and if so, whether an alternative energy program could achieve the same results at less cost.

The recent emphasis on energy for all often ignores the importance of a healthy debate such as the one in the 1980s. As so aptly put by George Santayana, "Those who cannot remember the past are condemned to repeat it." While energy for all is quite a good idea, ignoring or lacking understanding of the valid points debated in the past could end up doing more harm than good, perpetuating a cycle of unbridled optimism followed by excessive pessimism. Even today, the research of the 1980s contains many lessons for making the most of costly investments in rural electrification programs.

Improvements in rural infrastructure are often cited as preconditions for agricultural development. A distinction is usually made between physical and institutional infrastructure development programs; physical infrastructure refers to roads, rural electrification, large-scale irrigation and other similar projects, while institutional infrastructure refers to programs involving agricultural extension, disease, pest control, educational and research. The results in this chapter considered rural electrification as part of the total rural infrastructure. Its impact on agricultural development was examined in conjunction with other complementary programs, including credit, agricultural extension, rural market development and diesel irrigation pumpsets.

Rural electrification generally affects agricultural production in three distinct but related ways. First, electric-powered farm machinery (e.g., water pumps, fodder choppers and threshers) can function as direct inputs in the process of agricultural production. Second, electricity can have diffuse effects on agricultural modernization not directly captured in specific on-farm activities. Typical examples include electricity for cooling milk in collection centers, listening to radios, watching television or reading by means of better lighting. With radio or television, agriculturally relevant news and information can reach even the most remote audiences. Third, electricity can be used for lighting in on-farm operations, such as crop processing and poultry and dairy operations.

The 1980s surveys for the three countries in this book were specifically designed to assess the impact of rural electrification on development (Appendix). The results indicated that, in India, the impact was indirect through changes in irrigation patterns and agricultural innovations. But in Colombia and Indonesia, there was little impact on agricultural development. The conclusion was that supporting policies, including the type of agriculture and extent of a country's commitment to rural development, shape the impact of rural electrification on development.

Rural Electrification Development Model

The agricultural development model for rural electrification is based on the assumption that it is part of a land intensification program that includes irrigation (Hayami and Ruttan 1971, pp. 282–83); thus, the model applies mainly to countries at quite high levels of cultivation intensity and population density. The introduction of electricity in a region makes it possible for farmers to switch from traditional, and in many cases gravity-fed, irrigation sources. They also can change from diesel-engine pumpsets. Finally, farmers have the ability to irrigate farmland for the first time. With deep tubewells, submersible electric pumpsets may be the only practical means of efficiently lifting water to the surface. However, if the water table is relatively high, farmers can choose from electric pumps, diesel engines or traditional irrigation methods, such as the Persian wheel or leather bags drawn from a well.

Increasing the percentage of land under irrigation generally can affect agricultural practices by encouraging farm innovations. With irrigation of rainfed farmland, both double-cropping and changes in cropping patterns become possible. During India's monsoon season, for example, most agricultural production is rainfed, with irrigation pumping used only in areas of below-normal or sporadic rainfall. The benefits of irrigation typically are conditioned by level of rainfall, length of growing season, and availability of groundwater. During the winter season following the monsoons, the water table generally is high enough so that irrigation makes it possible to grow and harvest a second crop. Just as important, more remunerative crops can be grown in place of the traditional non-irrigated varieties. Superior cereals (rice and wheat) and cash crops (sugarcane and tobacco) can replace the inferior cereals (millet and barley) grown in most non-irrigated, dry areas. Thus, irrigation makes it possible to grow and harvest two or even three crops a year, producing significantly greater income for farmers.

The story of the Green Revolution is a well-known chapter in the history of development. In India, packages of fertilizers, hybrid seeds and pesticides were introduced to farmers in pilot programs during the 1960s. At the same time, the agricultural extension staffs in the regions were expanded to help disseminate information and advise farmers about the program. While substantial increases in agricultural productivity were realized, the pilot programs were unrepresentative; that is, they were limited to the most progressive districts. At the time, there was some concern that benefits would accrue only to the most advanced districts. But the program quickly spread to surrounding districts, and by the 1980s virtually all regions in India have access to hybrid seeds and other inputs.[2] The initially successful programs were those for wheat during the late 1960s, but hybrid rice varieties, mostly of the non-paddy variety, along with varieties of inferior cereal grains (e.g., sorghum and millet), became widespread during the 1970s. The productivity of the hybrid seeds, however, depended, to a large extent, on an assured and well-timed supply of water during certain critical periods of plant growth.

Figure 3.1 Model for Impact of Rural Electrification on Agricultural Development in India

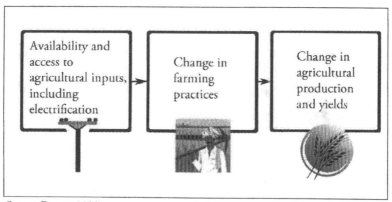

Source: Barnes 1988.

Rural electrification can indirectly cause changes in farming practices that result in higher agricultural yields (figure 3.1). Examples of such changes are increasing the percentage of village cropland under

---

[2] Brown (1971) provides a detailed analysis.

irrigation or stimulating agricultural innovation. Thus, the proposed model for rural electrification and agricultural development involves three stages: use of electricity, change in farming practices and increases in agricultural yields.

To evaluate rural electrification's role in the development process, it is necessary to analyze explicitly all of the complementary inputs and their relationship to rural electrification and changes in farming practices. Although this analysis is complex, the issues raised by critics necessitated controlling for access to electricity, agricultural extension, roads and other variables.

Changes in the Villages of India, 1966–80

I ndia actively pursued a policy of promoting electric irrigation pumps starting in 1966 because of serious droughts experienced in the previous decade. The dramatic growth in electricity for irrigation resulted from an explicit national commitment to improve agricultural productivity through rural electrification and substitute Indian coal for imported oil (table 3.1).

Table 3.1 Electric Agricultural Pumpset Growth in India, 1951–79

Period	Villages with electricity (no.)	Villages with electricity (%)	Electric pumps (no.)	Pumps per village with electricity (average no.)
1951	3,061	LT 1	--	--
1951–56	7,296	--	56,056	7.7
1956–61	21,750	3	198,984	9.1
1961–66	45,144	--	512,756	11.4
1966–69	73,732	13	1,088,824	14.8
1969–74	156,729	27	2,426,133	15.5
1974–79	232,042	43	3,599,328	15.5

Source: Rural Electrification Panel-Committee on Power 1979.

Note: All figures are for end of the plan period; (--) indicates data not available. LT stands for low-tension lines.

Consequently, the number of villages receiving electricity and the extent of irrigation through electric pumpsets increased dramatically in the late 1960s and early 1970s. The number of pumpsets per 1,000 hectares increased from approximately 2 in 1966 to more than 36 by 1977. During that period, the percentage of villages with electric pumps increased from about 12 percent to more than 40 percent.

The villages of India surveyed reflected this national commitment to rural electrification and expansion of irrigation. In the early 1960s, there were few pumpsets in the surveyed villages (see table 3.3). Most irrigation used traditional methods, including gravity-fed, animal and even human powered forms. Traditional well irrigation sources generally covered less than one acre and animal-driven irrigation provided only enough water to irrigate two acres.

The water output from diesel and electric pumpsets was far greater than from using traditional irrigation methods. Between 1966 and 1980, farmers bought both diesel and electric pumps to intensify cultivation. From 1951 to 1988, the average number of pumps per village surveyed increased from about 2 to more than 11 (figure 3.2).

Figure 3.2 Growth of Electricity Connections in India, 1951–81

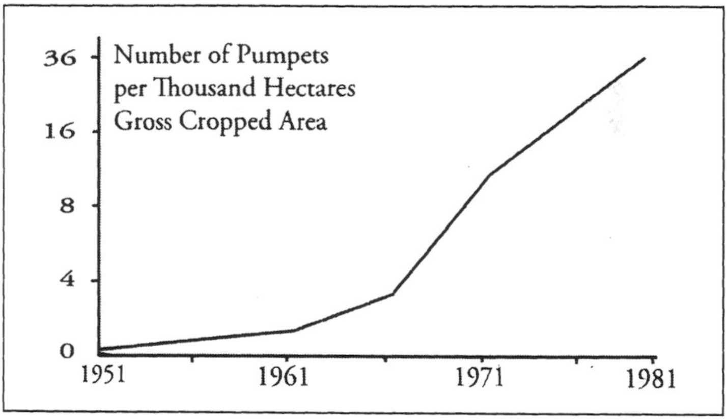

Source: India Survey 1981.

Because electric and diesel pumps are more efficient than traditional methods, there was a significant increase in irrigated farmland, which no doubt led to a change in cropping patterns. The amount of land under irrigation by pumps was manyfold greater than that using traditional irrigation sources (excluding canals). India's commitment to encouraging farmers to irrigate their land with diesel and electrically powered pumps in private wells was extremely successful.

Other important changes in the use of agricultural innovations occurred in the sample villages during the 1966–80 period (table 3.2). The average percentage of village leaders using fertilizers increased from 73 percent to more than 96 percent, indicating that fertilizer use was well established in nearly all of the villages sampled. Likewise, leaders' use of improved seeds increased quite dramatically, from 54 percent to 90 percent. Switching to new implements was more modest, rising from 35 to 48 percent.

Table 3.2 Village Leader Agricultural Innovations in India, 1966 and 1980

Village leader and new agricultural practices	1966	1980
Use (%)		
Fertilizers	73	96
Green manure	47	44
New implements	35	48
Improved seeds	54	90
Perception of utility (scale of 1–5)		
Fertilizers	4.0	4.3
Green manure	2.8	3.4
New implements	2.0	3.1
Compost pits	3.0	4.2
Improved seeds	3.3	4.5

Source: India Survey 1981.

Note: Figures are based on the same questions in the same villages for 1966 and 1980; new practices scale ranges from 1 = not useful to 5 = very useful.

By 1980, the villages had much greater access to agricultural inputs, including hybrid seeds, fertilizers and pesticides. This resulted from improved production and distribution systems, both private and public, which had eliminated many of the bottlenecks that existed in the 1960s. To measure access to agricultural inputs, the village leaders were asked: "Given the willingness to pay, how much difficulty have farmers had in acquiring key agricultural inputs?" From 1966 to 1980, village leaders reporting difficulty in obtaining fertilizers decreased by 21 percent on average (from 59 to 38 percent), while their difficulty in obtaining improved implements declined by 14 percent (from 56 to 42 percent).

Over that 14-year time period between surveys, the changes in India's rural infrastructure were a contributing factor to agricultural development (table 3.3). It was quite evident that infrastructure improved. Schools, transportation, mass media and banks all became more abundant in the countryside. The only exceptions were agricultural extension and rural markets. The market decline probably resulted from market consolidation, typical of agricultural commercialization. Because of improved transportation, local markets probably began to concentrate in central rather than dispersed locations. Such a process was typical for the development of growth centers and market towns (Wanmali 1983).

The decline in availability of agricultural extension services over that 14-year period is more puzzling since one would expect agricultural services to have improved. This trend requires some explanation. The sample contains 36 villages that were part of the Intensive Agricultural District Programme (IADP), which began in the early 1960s. These districts were singled out as being advanced and a good environment for inducing agricultural change through spurring the adoption of new agricultural technologies. Therefore, these districts received more agricultural services than the surrounding districts. This program was ended in the early 1970s. The program could have inflated the number of extension agents in the villages for the 1966 period, and returned them to more normal levels by 1980.

The villages in the survey represented a broad range of rural electrification. Some villages had had electricity for more than 15 years, while others had not yet received it. About 57 percent of the villages sampled had gained access to electricity service during the 14 years following 1966. In 1966, only 33 percent of the villages sampled had electricity compared to 71 percent by 1980. This is reflected in the number of years that the villages had electricity. In 1966, the average number of years with electricity was just over 1, as most villages had not yet been connected to the central grid. By 1980, the average number of years with electricity for the sample villages was over 8.

Table 3.3 Changes in Agricultural Practices and Infrastructure Proximity in Villages of India, 1966–80

Village characteristics	Year		Change
	1966	1980	
Irrigated area of net cultivated area (%)	20.3	22.8	2.1
Well irrigation of net cultivated area (%)	0.7	8.6	8.1
Multiple cropped area (%)	16.5	28.3	11.0
Agricultural innovation index (0-4)	2.2	2.7	0.5
Times talked to agricultural officer (0-24)	9.1	10.3	1.1
Literacy (%)	19.3	25.9	6.6
Village population (thousands)	1.6	2.0	0.4
Proximity to bank (0–4)	0.6	1.5	0.0
Schools above primary proximity (0–12)	4.2	4.6	0.4
Transportation proximity (0–12)	5.5	6.9	1.4
Mass media proximity (0–16)	5.4	7.4	2.0
Wholesale, retail market proximity (0–8)	3.5	2.3	1.2
Village grain mills (no.)	0.8	1.8	1.1
Electric pumps (no.)	1.2	7.6	6.1
Diesel pumps (no.)	1.3	6.0	4.7
Village years with electricity (years)	1.3	8.1	7.1

Source: Barnes and Binswanger 1986.
Note: Number ranges in parentheses indicate units for the 1966 and 1980 variables.

Credit apparently is quite an important input for agricultural development. The percentage of leaders indicating farmers had some difficulty in obtaining credit increased slightly between 1966 and 1980, but this might indicate that more farmers were trying to obtain credit in 1980 than in 1966. An average of about two banks per village were available for providing input loans, so credit was generally available in these villages. In fact, the village leaders estimated that 78 percent of the farmers who applied for credit had received it, but there was a wide variation in results.

Access to mass media measures the degree to which a village was connected to the larger society through the communications network (i.e., proximity to cinema, post office, telegraph, library and telephone). Between 1966 and 1980, mass-media access improved for all items except proximity to library. A "proximity to school" item measured village distance from middle schools and high schools. Primary schools were excluded from the survey because virtually all villages in the sample had them. As might be expected, only 15 percent of the villages had high schools, while 33 percent had middle schools.

Infrastructure Impact on Farm Production

Rural electrification, agricultural extension and credit programs may be locally available, but farmers must actually utilize these programs in order to realize an impact on their crop yields. During India's shift from traditional to more modern agricultural practices in the 1960s and 1970s, a period known as the Green Revolution, farmers' decision to irrigate was closely linked to modernizing their agricultural practices, including the use of hybrid seeds. The important changes in farming practices during the Green Revolution involved irrigation and multiple cropping, as well as agricultural innovations, which are reviewed below.

Irrigation and Multiple Cropping

The percentage of land under irrigation in villages directly results from the extent and type of irrigation. India has four main modes of irrigation: wells, tanks, rivers and canals. The means of irrigating land include electric motors, diesel engines, gravity and human or

animal power. The measures of irrigation used in this study are percent of land irrigated and percent of land multiple-cropped.[3] Generally, irrigated land is more valuable than non-irrigated land; likewise,
multiple-cropped land is more valuable than land irrigated for only
one season.

Diesel and electric pumps, both of which are more efficient and
convenient than traditional animal- and human-powered methods
of well irrigation, had become increasingly important in India by the
1980s. Canals systems, considered a modern method of irrigation,
were also important in many regions. In the 1980s, farmers were
abandoning the unwieldy animal-drawn leather bags and Persian
water-wheel pumps since these traditional irrigation methods provided little water to farmland. In 1966–80, the increase in electric
pumpsets per capita is associated with substantial growth in a village's irrigated area, from approximately 10 acres to 143 acres (table
3.4). The results over that period were similar for diesel-engine
pumpsets, with the village area irrigated reaching 138 acres. Thus, by
1980, both diesel and electricity figured prominently as energy
sources for irrigation.

Table 3.4 Village Area Irrigated by Source, 1966 and 1981

Irrigation sources	Irrigation sources per village (no.)		Total land irrigated per village (acres)
	1966	1980	1981
Electric pumps	2.1	11.6	143
Diesel pumps	2.7	13.6	138
Gravity-fed canals	--	33.3	27
Animal-lift wells	--	3.0	5

Source: India Surveys 1966 and 1981.
Note: (--) indicates data not available.

---

[3] For the purpose of this study, the percent of area irrigated is total land irrigated divided by cropped land (including area cropped for more than one season).
Multiple cropping is the percentage of land area farmed for more than one season.

The model of the impact of infrastructure on agricultural development involves the development of infrastructure, changes in farming practices and improved yields. The 1981 survey was able to model irrigation pumps, traditional irrigation, availability of credit and printed media as possible inputs for agricultural modernization. Results of the statistical modeling showed that the number of diesel and electric pumps influenced the percent of area irrigated in a village (figure 3.3). Diesel pumpsets and credit availability were influential in increasing the percent of land area farmed for more than one season.[4] In turn, all improvements in farming practices resulted in higher crop yields.

Figure 3.3 Diagram of Relationship between Infrastructure, Change in Farming Practices and Crop Yields in India, 1981

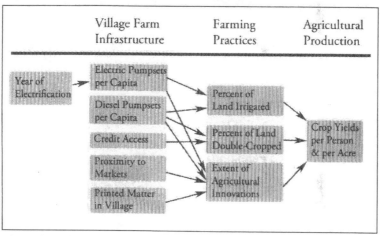

Source: India Survey 1981.

Note: The causality represented by the arrows resulted from a statistical model (Barnes 1988).

---

[4] The dependent variables in the model were percent of land irrigated and multiple-cropped, while the independent variables were year since electrification; electric pumps; diesel pumps; traditional irrigation; adequacy of credit; proximity to markets, schools, transport services and communication; newspapers; contact with agricultural extension and difficulty getting farm inputs.

Irrigation is virtually a necessity for farming cropland for more than one season, the exception being regions with high levels of rainfall, which can use residual moisture from the monsoon to grow an additional crop. Thus, during the 1980s, the percentage of village land irrigated and multiple-cropped were related because both needed a source of irrigation; however, diesel pumps were more strongly associated with multiple cropping, while electric pumps were important for the percentage of community land irrigated. Interestingly, credit availability was positively associated with multiple cropping, probably because farming land for more than one season required additional investments in seeds and fertilizer. But it apparently did not guarantee that farmers would be innovative. Irrigation pumping, access to markets and availability of print media were related to agricultural innovations, and credit appeared important mainly for farming land for more than one season. In short, the statistical modeling exercise showed that traditional irrigation methods were not strongly associated with the area irrigated or multiple cropping, reflecting the growing importance of commercial energy in rural India during the 1980s.

Agricultural Innovations

Agricultural innovations, including the use hybrid seeds, fertilizer and irrigation, were key components of the Green Revolution strategy. In this study, it was found that diesel and electric pumpset irrigation, market access and the reading of printed materials were related to a village leader's rating of agricultural innovation in the community.[5] Because water is extremely important for the success of new inputs, including hybrid seeds, the relationship of both diesel and electric pumps to innovations was positive, as expected; interesting-

---

[5] The dependent variable in the model was the index of village-level agricultural innovation, while the independent variables were year since electrification; electric pumps; diesel pumps; traditional irrigation; adequacy of credit; proximity to markets, schools, transport services and communication; newspapers; contact with agricultural extension and difficulty getting farm inputs.

ly, traditional sources of irrigation were not related to agricultural innovations. The existence of retail or wholesale markets in the vicinity of the village and reading or having locally available printed materials also were important positive factors for agricultural innovation.

India's rural electrification policy of encouraging investments in electric pumps appears to have had a substantial impact on farm practices in the countryside, but complementary inputs were also important. Certainly, diesel engines were a viable alternative power source for electric pumping. Credit and literacy were also vital for famers interested in innovative practices. In sum, rural electrification, combined with such factors as credit availability and literacy, played a significant role in helping to change India's farming practices.

Farming Practices and Agricultural Yields

For developing countries, yield per acre and yield per person are quite distinct concepts. Yield per acre, in many cases, reflects an intensification of traditional farming methods. As population density increases, farms may divide into smaller units, thus working the land harder to support more people (Geertz 1963). Yield per person is more of an indicator of capital improvements in agriculture. As increasingly more machines replace labor, yields per person are improved, but it is quite possible that yields per acre remain constant.

Three farming-practice variables could have a direct impact on crop yields: multiple cropping, irrigation and agricultural innovation. During the 1980s, agricultural innovations and multiple cropping both related positively to the indexes of agricultural production.[6] The second crop of the year generally was grown during the winter season, a period of scant rainfall when the water table was still high. Innovative farming practices had a positive impact on agricul-

---

[6] The dependent variables in the models were yield per acre and yield per capita, while the independent variables were percent of land irrigated, percent of land multiple-cropped and use of agricultural irrigation.

tural productivity. Villages with higher levels of innovation and multiple cropping had higher crop yields. Over the long term, a change in village farming practices led to an improvement in crop yields, which also could have had spillover effects on farm incomes and regional levels of development. It might be better to think of multiple cropping, innovation and irrigation as an interrelated set of farming practices associated with agricultural yields.

To summarize, this study found that rural electrification in India contributed significantly to village-level agricultural development. In the 1980s, the introduction of rural electrification and other complementary conditions had an impact on irrigation and agricultural innovation. In turn, this interrelated set of farming practices was related to agricultural yields. Thus, rural electrification was not the only factor affecting farming practices; other complementary factors included the existence of rural markets, credit, diesel pumps and access to printed information.

During India' Green Revolution, rural electrification played an important role in stimulating agricultural productivity. But without the complementary emphasis on water pumping and use of hybrid seeds and other new agricultural technologies, rural electrification would have had much less chance of working, as was the case in Indonesia and Colombia.

Growth of Electric Agricultural Pumps

It is quite important for the agencies providing rural electrification to be able to forecast the expected connection rates and thus the cash flows that will be derived from extending electricity to particular types of villages. Evidence from the 1980s suggested that rural electrification was related to agricultural productivity, but this could have been a two-way relationship. That is, wealthy villages with high levels of agricultural production may have been better able to bear the financial burden of purchasing pumps, digging wells and paying for the operating costs of pumps. A close examination of patterns of electric pumpset growth helped to unravel whether agricultural innovations were causing the growth of pumpsets or vice versa.

In the agricultural sector, various factors affected the growth of electric pumps. Some farmers may have wished to wait and see whether the capital investments of other farmers would have significant benefits before adopting an electrical pumpset themselves. Also, installation of an electric pumpset may have meant abandoning an already used method of irrigation, such as a diesel engine, Persian wheel or canal. Irrigation investments might have entailed additional costs, including field leveling and construction of irrigation ditches. Adopting an electric pump might have meant dismantling, selling or simply not utilizing an existing irrigation method. To adopt electric pumps, farmers had to be willing to abandon their current irrigation practices or otherwise wait for their existing systems to depreciate physically or become highly uneconomical.

The growth rates for agricultural pumpsets and electrical connections confirm that investment in electric pumpsets is a slow, cumulative process. The line in figure 3.4 represents the average number of electric pumpsets by year of electrification for villages with electricity. Only those villages having agricultural connections have been included, on the assumption that those without electric pumpsets either do not yet have electricity or may have other constraints, such as a low water table. For the 76 villages in the study with agricultural connections, the mean growth rate of connections of pumps per year is 1.24. This figure indicates a slow but steady increase in agricultural pumpsets after village electrification; that is, villages with a longer history of electrification have more electric pumpsets.

The main factor that affected the growth of electric irrigation pumpsets was the year that a community received electricity.[7] Interestingly, regions that started out in 1966 with a significant amount of

---

[7] A regression analysis was used to examine the relationship between initial village conditions in 1966 and the number of agricultural connections for villages in 1980. The dependent variable was 1980 village electricity pumpsets; the independent variables were years of village electricity and the 1966 levels of crop yields, agricultural innovations, multiple-cropped area, human/land ratio and diesel pumps.

double-cropped land had a lower than average number of agricultural pumpsets by 1980, perhaps because they already had canal irrigation. This means that regions without already well-developed irrigation networks were better environments for electric pumpset growth. Alternative irrigation systems, such as canals or diesel engines, are a hindrance to the adoption of electric agricultural pumpsets. In promoting electric pumps for agricultural use, planners should thus avoid villages with extensive irrigation systems already in place. They should also be concerned with whether adequate credit is available for investment in pumpsets and wells.

Figure 3.4 Growth of Electric Pumpsets in India, 1955–80

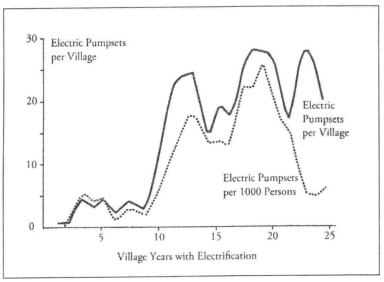

Source: India Survey 1981.

Interestingly, agricultural innovations in 1966 were not related to the 1980 levels for agricultural pumpsets. Since innovations would not seem to be a long-term cause of growth in agricultural connections—but they are related by 1980—one could speculate that high levels of agricultural innovations may have been caused by rural electrification.[8] When farmers were asked in 1980 what source of energy they preferred for irrigation, more than 90 percent said they preferred electricity for agricultural pumping. Their preference for electricity may have been caused by a combination of factors, such as a desire for improved agricultural production, subsidized electricity prices for agriculture or the minimal maintenance required by electric pumps compared to the diesel alternative.

Conclusion

The India study indicated that the productive impact of rural electrification is to promote agricultural pumping and, to some extent, agricultural innovations. Agricultural innovations and double-cropping, in turn, are associated with improvement in agricultural yields, all patterns which are still common today (Monari and Mostefai 2001). The implication was that rural electrification, through its impact on agricultural pumping and use of agricultural innovations, had net positive effects on agricultural development in villages. India's policy of subsidizing rural electrification during the 1980s seems to have had a positive impact on agriculture. Although it is probably true that electricity from coal and hydro-power was being substituted for diesel power, diesel engines were still quite prevalent in rural India, and were positively associated with factors leading to improved agricultural yields. Today most farmers in India use electric pumps, and irrigation with diesel engines is limited to only the most remote farms.

Some interesting observations can be made about government policy and the agricultural impact of rural electrification in India as compared to Colombia and Indonesia. Neither Colombia nor Indonesia had attempted to improve rural productivity through rural electrification used for either agricultural pumping or driving other farm machinery. In contrast to India, neither country had programs to promote electricity use in agriculture. In Colombia, few farmers

---

[8] Barnes and Binswanger (1986) provide a more rigorous panel analysis.

had pumps for irrigation, and use of improved seed varieties was negligible. In Indonesia, some tobacco farmers were unaware that electricity could be used for irrigation pumping and expressed a desire for diesel pumps. Of course, at the time the Indonesia rural electrification schemes had only been in operation for a short time, so there may not have been enough time for farmers to invest in electric pumps and other machinery.

The positive impacts of rural electrification on agriculture in India, and perhaps the lack of impact in Colombia and Indonesia, most likely were determined by government policies and complementary development programs. These programs would include a variety of improved agricultural inputs that combined effectively with unique regional characteristics. Obviously, the irrigation approach adopted by India is not applicable to all countries because of climatic and geographical differences. Rural electrification programs offer substantial opportunities for agricultural development, but such opportunities for productive uses may be hampered by a failure to implement complementary programs. Even today increasing the productive impact of electricity for agriculture in most countries has not been emphasized enough, probably because it requires more concerted effort to coordinate rural electrification with other relevant programs.

4. Small-Scale Industry and Commerce

Industrial development today is almost inconceivable without electricity. In industrialized and developing societies alike, electricity is the energy that drives the great maize of industrial machines. It would seem obvious that electricity is necessary for the development of industry. But the central question is whether rural electrification will bring about the development of decentralized industries so that economic benefits will accrue to people living in rural areas. Will new industry and commerce relocate to or develop in rural areas once they receive electricity? Are rural industries that already have electricity more productive than unelectrified rural industries?

Advocates of rural electrification contend that electrification is a necessary precondition for the development of rural industry and commerce and that it will provide employment (Dinkelman 2011; Grogan and Sadanand 2012), improve rural incomes and lead to growth in rural areas, which will counteract the present trend of massive urban growth. According to their view, rural industries stimulate the growth of small towns and centers distant from the major cities, bringing essential services to rural areas. These predictions parallel some interesting observations made in a 1925 essay by

Ernst Bradford on the potential impact of electrification on factory location in America:

> The effect of cheap [central grid] power...is likely to add somewhat to the movement of factories away from the great cities and to change the map of our industrial power-using centers by adding to the number of small- and medium-sized manufacturing cities and towns....The massing of population in urban centers, which is partly a result of the urban locations of factories, will also be affected by cheap power. Central grid power will operate almost certainly as a force making for decentralization. With the factories as they move to the country will naturally move also the workers who formerly clustered at the mill gates in the city, and with this spreading out of population will disappear some of the ills that come from having too many families per acre of ground....(Ernst Bradford, "The Influence of Cheap Power on Factory Location and on Farming," *The Annals of the American Academy of Political and Social Science*, vol. 118 [March 1925], 91–95)

Today advocates would add that electricity will also stimulate the development of new industries in rural areas (Cabraal, Barnes, and Agarwal 2005).

The critics counter that the record to date of small-scale industrial and commercial development in rural areas with electricity has been poor; the growth of rural industries in developing countries simply may not parallel the experience of the United States or other now developed nations. Electricity has not produced the bonanza of growth conceived by the rural electrification planners. The industries that have been established are predominantly grain-milling units, which could be powered by diesel engines rather than electricity from a centralized grid. According to critics, industry will not locate in rural areas until such problems as inadequate transport and communications facilities are alleviated. Larger towns and factories can be provided with electricity without placing the nation and its utilities under the financial burden of covering the whole country with an expensive power-distribution system. Alternative energy programs for rural areas may be a more appropriate and less expensive energy strategy.

However, by failing to extend electricity to rural areas, much development potential may be lost. Mellor (1976) argued for an employment-oriented decentralized growth strategy for developing nations, premised on large investments in rural infrastructure, including electricity. The development of large-scale urban factories without comparable investments in rural areas might result in the uneven development patterns typified by the so-called dual economies observed in many Third World countries, where the rich get richer and the poor get poorer. According to Mellor, direct investment in public services that facilitate small-scale industrial and commercial growth should take precedence over the construction of large-scale factories.

This chapter presents the conclusions from the 1980s surveys conducted in India, Colombia and Indonesia. The conditions conducive to the growth of industry and commerce included a combination of electricity availability, adequate markets, available credit and a high rural literacy rate.

Profile of Rural Businesses

Rural commerce and industry in India, Colombia and Indonesia are typical of those in many areas of the developing world; food processing, commercial shops, and repair shops were the principal businesses identified in the study (table 4.1). Of the three countries, Indonesia had a greater diversity of shops and manufacturing firms in the 1980s, reflecting the high level of development in the specific region sampled. Of the three studies, Indonesia had the most comprehensive business sample, which included a census of all 131 businesses, comprising manufacturing, services, food and commercial enterprises. Examples of the types of businesses were furniture making, rice milling, sawmills and general stores. Nearly all of the businesses in the sample had adopted electricity when it became available in the community about three-to-four years before the survey. Therefore, the Indonesia analysis can best be described as an assessment of the short-term benefits of electrification for a relatively advanced region. However, these short-term benefits may persist over time and may well turn out to be long-term benefits.

The India study surveyed industrial establishments, but did not include the country's traditional caste/artisan occupations. Blacksmiths, potters, weavers, carpenters and leather workers comprised the caste occupations typical of Indian villages; these traditional occupations, which in many instances were single-family businesses, had been the focus of many ethnographic studies. The Indian survey covered flour mills, food services, bakeries, sawmills and other industrial and commercial units. About nine out of ten of the industries sampled in India were grain-processing firms.

Table 4.1 Rural Business Types in India, Indonesia and Colombia, 1981

Industry type	India	Indonesia	Colombia
Agricultural processing	87	24	--
Commercial establishments	3	30	63
Rural industry	5	30	37
Rural services	5	16	--

Sources: India Survey 1981; Indonesia Survey 1981; Colombia Survey 1981.
Note: (--) indicates data not available; business-type categories differ slightly by country.

The 1980s household survey in Colombia found that many households owned some type of retail business or shop. The majority of businesses were commercial shops selling such goods as cigarettes, sodas, beer, bread, food and kerosene. There were a significant number of specialty shops in the communities, selling such items as baked goods, ice and work clothes. A small percentage of the sample businesses were engaged in artisan activities and repairs.

Business Development Patterns in India and Colombia

Many early studies on the impact of electrification on rural businesses have been criticized for surveying only existing industries, thus leaving out those that have gone out of existence. As a way of circumventing this difficulty, the 1988 study for India and Colombia included the year in which existing businesses were established in order to determine the rate of business establishment for particular time periods (Barnes 1988). For Indonesia, patterns of business development were not examined because of the short period that communities had had electricity.

The development of small businesses should be viewed as a natural part of the dynamic development process. Typically, developing rural businesses are extremely small; many are family-run establishments that change owners and sometimes go out of business within just a few years. Most businesses in India and Colombia were started 5–10 years before the survey, suggesting, as expected, that there was a high turnover rate of rural businesses (figure 4.1), as is also the case in developed countries. In Colombia, the percentage of businesses started during the 5 years before the survey was even higher than for India. Remarkably, 15 percent of businesses in Colombia and India have survived for 15 years or longer.

Figure 4.1 Business Establishment by Year in India and Colombia, 1981

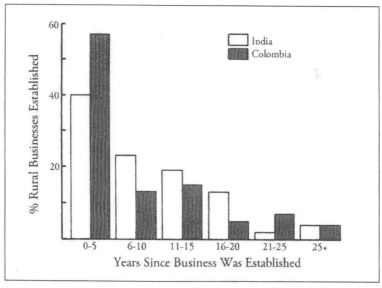

Sources: India Survey 1981; Colombia Survey 1981.

In India, the start-up year for industries is related to the year of village electrification. As illustrated in figure 4.2, when the year of electrification is controlled for, the post-electrification era for the villages has a higher level of business activity. More than 38 percent of the village industries still operating in 1980 were established during the first 5 years following electrification. Even after the fifth year of electrification, there was still a higher level commencement of businesses than during the pre-electrification period.

Figure 4.2 Business Establishment Before and After Electrification in India, 1981

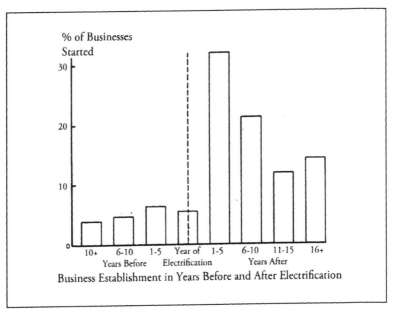

Business Establishment in Years Before and After Electrification

Source: India Survey 1981.

The increase in business activities is also reflected in village employment patterns. By 1971, about half of villages in the India sample had received electricity. That year, 6 percent of workers were employed in industry in the village sample with electricity, as compared to 3 percent in villages without electricity. Household manufacturing also proved to be positively associated with electrification, suggesting that the advent of new industries in the villages had not negatively affected home industries. Only in the Punjab, which is characterized by more specialized industries, were there fewer people employed in household industries within villages with electricity. The conclusion is that rural electrification in India has stimulated the development of rural industry.

For Colombia, the relationship between electrification and business activity is similar to that of India. Even though its absolute level of business activity is somewhat low, the longer a community has had electricity, the higher its level of business activity (figure 4.3). This means that the absolute level of industry and commerce is positively associated with year of electrification.

The percentage of households with a business in Colombia is higher in villages with electricity and tends to increase over time (table 4.2). Finally, nearly all household businesses were started after their respective communities received electricity. The study findings for Colombia reinforce the India findings and those of several studies in the Philippines (Costas 1982; Madigan 1981, pp. 12–14).

Figure 4.3 Village and Commerce Growth by Year of Electrification in Colombia, 1981

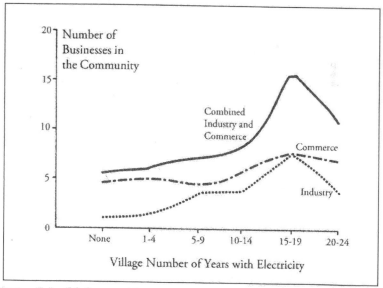

Source: Colombia Survey 1981.

Table 4.2 Increase in Household Rural Businesses Before and After Electrification in Colombia, 1981

Years before or after community received electricity	Percent of existing community businesses started
11–15 before	10
6–10 before	12
0–5 before	14
0–5 after	16
6–10 after	17
11–15 after	19
16+ after	29

Source: Colombia Survey 1981.

The response of business to electrification appears to be quick and dramatic for rural industries. When villages in India obtained electricity, 92 percent of industries commenced operations with electricity (table 4.3); as more villages received electricity, more industries started up using it. During the 1961–65 period, only 7 out of 10 industries began operations with electricity, compared to 9.6 out of 10.0 during the 1976–80 period, when electrification programs were more intensive. Thus, it is virtually guaranteed that, in villages with electricity in India, any new business will start with electricity. Likewise, in villages with electricity in Colombia, 9 out of 10 businesses starting up adopted electricity. The same pattern was found in Indonesia, where most businesses adopted electricity within a year after the region came under the rural electrification program.

The evidence is quite persuasive that business development is enhanced by electrification programs. Although the number of rural industries per village is still low in India and Colombia, rural businesses do appear to be taking advantage of electricity for lighting and other end uses. In Indonesia, nearly all businesses use electricity for lighting, but electricity's replacement of diesel engines for driveshaft power is less common, primarily because of the country's heavy diesel subsidies.

Table 4.3 Percent Existing Industries after 1981 Community Electrification in India

Period after community received electricity	Percent of existing community businesses started
First 6 months	90
Second 6 months	2
2nd year	3
3rd year	0
4th year	1
5th year	2
6+ years	2

Source: India Survey of 128 Industries 1981.

Village and Firm Comparisons

Shops can use electric lighting to extend the business day, and mills can exploit electric motors for processing grains. Although this value of electric power for industry is well known, few comparisons have been made of firms with and without electricity. The benefits of electricity should be easy to observe and identify for rural firms.

It has been established that, in villages with electricity, nearly all shops and industries adopt it once available; in addition, there are more businesses in villages with electricity as compared to those without electricity. To further evaluate the effect of electrification on businesses, it is necessary to analyze the benefits of electrification at the firm level. For India, the profile of firms with and without electricity was examined for both household and non-household industries. For Colombia and Indonesia, owners of firms with and without electricity were asked questions on the utility of electricity.

Small-Industry and Household Manufacturing in India

Village leaders in the India survey were asked the question, "What has been the consequence of rural electrification on industrial growth?" For 7 out of 10 respondents, the answer was that the impact had been significant. Although the number of industries in Indian villages was still somewhat low, the leaders' subjective impression was verified by examining the differences between firms in villages with and without electricity. Villages with electricity tended to have more industries and more industrial employees, and firms had lower fuel costs, higher labor productivity and a greater diversity of processed or manufactured products (Samanta and Sundaram 1983, p. 142).

Direct comparisons between firms with and without electricity are necessary to clarify some of the benefits of electricity (table 4.4). On average, the capital size of firms with electricity was nearly twice as large as that of firms without electricity, reflecting greater investment in equipment. Because of the greater size, more workers were employed. Labor productivity was higher, probably because capital was substituted for labor in firms with electricity, so the machines improved the productivity per worker. Fuel costs as a percent of total firm costs were lower despite the fact that electricity rates for rural industries were three times higher than those for agricultural pumping, which was heavily subsidized. Smaller firms without electricity tended to use less fuel than larger firms, but apparently were less efficient in their energy use. For smaller firms without electricity, fuel expenses accounted for more than two-thirds of their overall cost of operation. Also, the products of firms with electricity tended to be somewhat more diversified. Thus, the village-level findings were confirmed by the comparison of industries with and without electricity.

Household manufacturing or cottage industries are sometimes considered in competition with commercial industry. One classic example is from the late 19th century when India exported cotton to Great Britain. The English mills manufactured the cotton into cloth and then returned the finished product to India, where it enjoyed a substantial price advantage over domestically produced cloth. Household cloth manufacturers in India were virtually eliminated.

Table 4.4 Industries with and without Electricity in India, 1981

	Average for industries	
	With	Without
Business characteristics	electricity	electricity
Size of industrial capital (Rs.)	39.5	18.1
No. of full-time employees	2.7	2.1
No. of part-time employees	0.2	0.1
Capital productivity	0.8	1.0
Labor productivity (Rs.)	10.9	8.3
Capacity utilization (%)	31	33
Fuel cost to total (%)	53	69
Net return on gross fixed assets	44	32
Diversity coefficient	2.5	1.6
No. of industries	122	21

Source: India Survey 1981.

Note: Capital productivity is the ratio of gross sales to gross fixed assets. Labor productivity is gross sales (in Rs.) per unit of labor input. The diversity coefficient is the ratio of all products manufactured across all units divided by the total number of units.

There is concern that introducing more capital-intensive industries with electricity in villages might eliminate labor-intensive cottage industries. While this may occur over the long run, the comparisons made between cottage industries with and without electricity reveal that electricity seems to benefit household manufacturing. Although the India survey had only a small number of cottage industries, most of those with electricity had extended hours and improved income for their rural households. Superior lighting rather than use of appliances or power tools was the major impact of electrification for home industries. Thus, there was no demise of cottage industries for the Indian villages.

Perceived Business Benefits in Indonesia and Colombia

Comparisons between businesses with and without electricity were not possible for Indonesia and Colombia. Instead, business owners were asked a series of questions concerning their perceptions of electrification's benefits. Most businesses in the Indonesia sample had received electricity during the previous year and a half, so the owners should have been able to evaluate the short-run impact of electrification. Since most of the businesses in Colombia had been recently established, owners could easily reflect on the difficulties they might have encountered starting their businesses without electricity.

Businesses develop and improve for many reasons. In Indonesia, the owners were asked whether electricity had provided the basis for business development (table 4.5). In Colombia, the question specifically referred to how the businesses had developed (table 4.6). About half of the businesses in Indonesia indicated they had "developed" due to electricity. Just less than half of the sample in Colombia reported improved sales and business conditions, and a significant percentage of the businesses were using new equipment. The results indicated that some new businesses had developed over and above what would have existed without having introduced electricity in the area. However, are these figures for Colombia and Indonesia high or low? After all, they show that about half of the businesses did not develop with electrification. Access to electricity is not the only factor necessary for developing a viable business sector. Market conditions, access to credit, relative prices of the products sold and other factors can affect business improvement or decline.

The specific end uses of electricity for business development included lighting, refrigeration, motors and ovens. In Indonesia, the various types of businesses used electricity in quite different ways. Service and manufacturing industries made use of electric appliances; commercial establishments benefited from indirect community changes, such as improved security and economic conditions, and food shops used electricity primarily for lighting.

Table 4.5 Perceived Benefits of Electricity in Indonesia, 1981

Perceived benefits of electricity for businesses with electricity	Percent of businesses
Better, more convenient lights	71
Cheaper than other fuels	24
Extended work hours	19
New machinery possible	16
Sample no. of businesses = 123	

Source: Indonesia Survey 1981.

The electrification of businesses in Indonesia did not lead to a significant increase in employment. The use of electric lights led to an extension of work hours in businesses with electricity (Brodman 1982, pp. 36–39), but this apparently did not result in any extensive hiring of new employees. Instead, existing employees were asked to work longer hours. Employer-employee relations in these villages involved informal ties, which often included providing financial assistance when the employee was sick and the giving of gifts at holidays. By the same token, the employee often had to assume responsibility for running the business when the owner was absent from the workplace. This mutual relationship led to employer-employee confidence and trust. By working longer hours, current employees were compensated by an increase in their salaries. As a result, the employer would not lay off workers in bad economic times and also would not expand or hire new workers during good times.

Thus, in Indonesia, the operation of machines or appliances by the small number of employees had not led to new employment or higher hourly pay, although productivity had likely increased. At the same time, the introduction of machines in businesses had not caused any decline in the number of persons employed. Apparently, the employers considered electrical equipment a means to improving productivity and product quality, without giving much thought to cutting back on present employment.

In the evaluation of productive uses of electricity in Colombia, the survey respondents identified activities that would have to be eliminated if electricity were not available. Once again, lighting proved important, as did a possible reduction in work hours and refrigeration. In all, two-thirds of the Colombia sample indicated that, without electricity, they would experience some difficulty in running or maintaining their businesses.

Table 4.6 Perceived Benefits of Electricity in Colombia, 1981

Perceived problems of not having electricity for businesses with electricity	Percent of businesses
Work shifts reduced	29
Refrigeration eliminated	23
Other	11
None	33
Sample no. of businesses = 107	
Anticipated benefits of adopting electricity for businesses without electricity	
Lighting	55
Refrigeration	56
Use of motors	16
Fans	14
Electric ovens or stoves	7
Sample no. of businesses = 20	

Source: Colombia Survey 1981.

In both Indonesia and Colombia, business owners clearly perceived the value of electricity. The electric appliances that would benefit businesses included lighting, refrigeration, motors and fans. The anticipated benefits for businesses were increased work shifts and extended business hours, mostly due to improved lighting.

Growth of Electricity-Using Businesses

The location of villages in a service economy might be a factor in the development of small businesses. In her classic work on the development of two villages in a recently irrigated region of South India, Epstein (1973) found that the village with newly irrigated land became economically more dependent on agriculture. By contrast, the village in the non-irrigated area became a service center, acting as a connection between the agricultural village and the larger society. Wanmali (1983) found that, as a region advances, a hierarchy of service centers develops, including those providing higher- and lower-order services. This type of specialization may result in an increase of pumps and agriculture-related items in some villages, while others may become commercial centers. Thus, village conditions for the growth of agricultural pumpsets appear quite different from those necessary for the development of business and commerce.

The approach used to analyze the growth potential of electricity-using businesses in India differed from that used for Indonesia. The Indonesia study examined the complementary conditions through interviews with businesses that had been able to use electricity productively. The India study covered the factors associated with greater-than-average growth of business electrical connections in villages with different socioeconomic profiles. Thus, the Indonesia study compared business conditions, while the India study focused on comparisons among villages. Unfortunately, for Colombia, no information was available on the conditions for growth of businesses.

Complementary Conditions in Indonesia

Many of the businesses in Indonesia during the 1980s used electricity productively, but there still was tremendous potential to enhance such use (Brodman 1982). High tariffs and costly installation charges for business electric driveshaft power discouraged electricity use. Of the 27 businesses in the region that could use either electricity or diesel fuel for driveshaft power, 18 reported that high electric installation costs made diesel engines preferable to electric motors.

The importance of electricity pricing was highlighted in the survey responses of grain-mill owners and operators in India and Indonesia. In India—an oil-importing country where diesel fuel is heavily taxed—most grain mills immediately switched from diesel to electricity. In Indonesia—an oil-exporting nation where diesel is heavily subsidized—grain mills did not adopt electricity for motive power. Yet more than half of service providers and manufacturers perceived electricity was advantageous, even at the high price (table 4.7). Thus, in Indonesia, the productive uses of electricity were for lighting, appliances and hand tools—a situation that was not likely to change until subsidies were eliminated for diesel fuel.

Table 4.7 Business Leaders' Perception of Electricity in Indonesia, 1981

Business type	Perception that electricity is advantageous	
	Percent agreeing	No. of sample businesses
Services	67	18
Manufacturing	51	45
Commercial	30	30
Agriculture and food processing	10	30

Source: Indonesia Survey 1981.

Business development in Indonesia's villages was also inhibited by capital scarcity, poor product marketing channels and inadequate information on electricity (table 4.8). While smaller businesses were able to increase profits from the use of electricity at about the same rate as larger firms, they lacked the same access to capital in the form of bank credit.

During the 1980s, a problem with existing credit programs for small and medium businesses was that most had been in existence prior to electrification of the region. In Indonesia, most credit programs were oriented toward agriculture, and land was a collateral requirement for most loans. As a result, small businesses did not qualify. This situation highlights the fact that introducing electricity to an area may alter predominantly agricultural rural credit needs to ones that are more business-oriented. Marketing products is another limitation endemic to many small businesses, with or without electricity. Market opportunities for rural businesses centered on agricultural goods and services and household domestic needs. In some instances, such opportunities could have been improved through cooperative marketing schemes. Typically, success in marketing was related to health of the local rural economy, without which the best marketing schemes were doomed to failure.

Table 4.8 Business Leaders' Reasons for Not Using Electricity Equipment in Indonesia, 1981

Reason given for not using electricity equipment (%)	Business size (%)		
	Small	Medium	Large
Lack of market demand	60	22	31
Electricity or equipment expense	40	64	38
Lack of information	0	14	23
No access to formal credit	89	85	53

Source: Indonesia Survey 1981.

Rural electrification programs in the United States were accompanied by intensive information campaigns to raise public awareness of the potential uses of electricity. In Indonesia, by contrast, businesses lacked any information on the comparative costs or maintenance required for electric motors as compared to diesel engines. Many large businesses were uninformed or misinformed about the potential uses of electricity. The relative newness of the electrification program in Indonesia might have been responsible for the inadequate marketing of the benefits and/or drawbacks of electricity. In retrospect, information dissemination might have been a cost-effective way to stimulate electricity use and rural production.

Two survey results from the Klaten area of Indonesia highlighted the need for complementary programs. Many businesses were using diesel engines for driveshaft power even though electricity was available, and some businesses indicated they would have liked to use electricity productively but could not obtain the credit to buy the equipment. For Indonesia, the electrification schemes should be coordinated with credit, marketing, and/or information programs that address related needs of electricity-using businesses. Also, pricing policies for both diesel fuel and electricity need to be examined to determine what price levels would be most beneficial for the general economy.

Business Connection Growth in India

Many past studies on rural electrification and business development in India included an analysis of "central places" or growth centers (Sen et al. 1971). Village industry and commerce tend to grow the fastest in villages servicing surrounding villages (Small Industry Extension Training Institute 1976). Business development may be retarded by the absence of local entrepreneurship, delays in sanctioning connections, lack of coordination between the utilities and other agencies and/or an irregular supply of electricity (Jain 1975). Other constraining factors to rural industry growth include inadequate capital, lack of credit and uncertainty about sales and markets (Center for Studies of Small Scale Industries 1979). Human capital also may be essential for business development. The classic work by Schultz (1964) stressed the importance of education for agricultural development. Educated business owners may be more capable of operating and expanding their business, and an educated population may be a more favorable market for business products.

The 1980s analysis of business connections in India revealed that both human and physical capital are important for growth of rural businesses. Villages with a longer history of electrification had a higher level of industry and commerce (figure 4.4). After villages received electricity, businesses obtained an electrical connection almost immediately. In this broad cross-section of villages at varying levels of socioeconomic development, only a limited number were growth centers. Nevertheless, the average number of commercial establishments was about 5 per village, accompanied by 1.5 industries. However, electricity alone did not cause dramatic numbers of new businesses to be established. Once most of the existing businesses had adopted electricity, the number of new connections slowed significantly simply because few businesses, and especially industries, were starting in the villages. The slow development of business connections during the 1980s differed substantially from the steady growth in agricultural connections described in the previous chapter.

Figure 4.4 Commercial and Industrial Connections by Years with Electricity in India, 1981

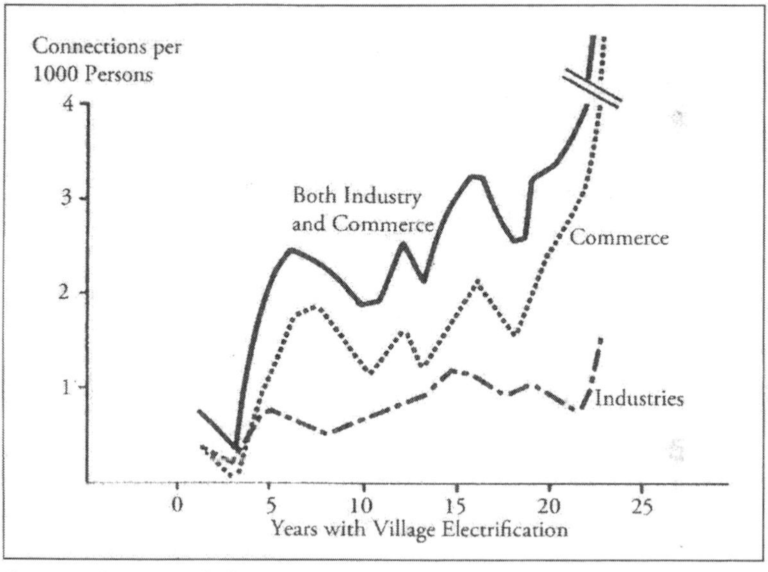

Source: India Survey 1981.

Village population size obviously was a major determinant of the number of industrial and commercial electricity connections. Very small villages could not support a large number of businesses simply because there were not enough people in the surrounding area to buy the products or services. Without markets, businesses obviously cannot grow in number. The findings showed that the factors contributing to village business growth were adequate credit, high levels of literacy and year of community electrification (Barnes 1988). The conclusions for India and Indonesia were remarkably similar despite quite different study designs. Adequate credit and rural markets were found to be essential complementary conditions to rural electrification for the establishment of viable businesses.

Conclusion

Simply stated, rural electrification has a significant impact on rural industry and commerce. Firms in rural areas generally are small, with a high turnover rate; they require every advantage in order to stay in operation. The number of businesses was generally higher in rural areas with electricity in India, Colombia and Indonesia. Comparisons between regions with and without electricity revealed that there were systematic increases in the number of businesses in the regions with electricity. The types of industries were typical of those found in most developing nations. They included grain mills, saw mills, furniture makers and general stores. In all three countries, nearly all businesses adopted electricity once it became available. In addition, virtually all new enterprises adopted electricity within their first year of operation.

One consequence of having a higher level of industrial and commercial development was greater nonagricultural employment. One might expect that investments in electrical equipment would have led to fewer workers because capital, in the form of machines, would substitute for labor. Although labor productivity was higher for firms with electricity, there was no evidence that machines were putting employees out of work. To the contrary, regions with electricity had a higher demand for business or industrial labor. Since the Indonesia study covered a relatively short time period, longer-term patterns for that country may turn out to be similar to those for India and Colombia. In Indonesia, the higher levels of production did not result in firms hiring new employees; rather, businesses extended the paid hours of existing workers. In India and Colombia, there were greater numbers of business and industrial workers in regions with electricity.

Household manufacturing was not hampered by rural electrification. If anything, the use of electric lights may have improved productivity in cottage industries. Certainly, it was true that many household industries went out of existence because of competition from urban-based industry. Village potters could no longer compete with the mass-produced pots and buckets from factories in urban centers. But the difficulties of household industry and artisans did not result from competition from local rural industries, which were predominantly agricultural or service-oriented. Local rural industries stimulated by rural electrification did not compete with household manufacturing. Most household industries did not use electric hand tools, and the primary benefit of electricity was improved household lighting.

The fuel-substitution patterns found in India and Indonesia highlighted the importance of energy prices. As an oil-importing country, India maintained a fairly high price for petroleum products, while Indonesia, an oil-exporting nation, heavily subsidized diesel power. In India, electricity was not as heavily subsidized for industry as it was for agricultural pumping; however, it was used for most rural driveshaft power applications, including grain mills. By contrast, in Indonesia, grain-mill and industrial owners with diesel engines did not purchase electric motors. Similarly, a study in Ecuador—another oil-producing country with extremely low-priced diesel—found that electricity was not used for driveshaft power in agro-industries (Kessler et al. 1981). Thus, diesel fuel subsidies discourage the use of electricity for driveshaft power applications.

Even though electricity appears to have improved levels of rural industry, their levels remained low. The average number of rural industries in India was only 2 per village with electricity, with only somewhat higher levels in Indonesia and Colombia. For Indonesia, the study focused on a more advanced region, which may explain its higher level of business activity. The expectation that electrification would lead to an explosion of business activities in rural areas remained unfulfilled. Nevertheless, electrification of regions does encourage additional business development and improves rural production.

The complementary programs necessary to stimulate rural business development were extremely important for rural businesses. The failure of many small businesses apparently resulted from internal business problems, including poor management or low educational level of the business leader. Constraints to business growth were more related to the lack of other rural infrastructure, such as rural roads, market conditions and access to capital. The India and Indonesia studies strongly support the notion that sufficient capital and adequate markets were necessary for the development of small businesses. Rural electrification programs that do not include or evaluate markets and credit availability are likely to witness slower-than-average business growth. A shortage of credit holds back business growth because rural borrowers lack the necessary start-up or expansion capital. However, growth centers will evolve in agriculturally advanced areas, and programs to facilitate small-scale business would be conducive to growth in these regions.

Human capital also plays an important role in the establishment of rural business. A higher level of literacy in a region translates into entrepreneurs' founding new businesses and better managing their businesses once established. Also, an educated population may be more likely to purchase appliances and other goods and services, thus providing a better market for businesses. The important role of human capital in business development emphasizes the need for a more critical examination of rural electrification's social impact, which is the focus of the next chapter.

5. Rural Social Change

The arguments in favor of rural electrification have often centered on its potentially powerful transformative effect on rural households. Imagine two rural households, one with electricity and the other without. In the evening, the household with electricity is a beehive of activity. Children are reading, the mother is busy mending clothes, and the father is listening to a radio program. By contrast, the household without electricity has a fairly unproductive evening and retires early, given that kerosene lanterns or candles emit dim light inadequate for reading or such close work as sewing. While these contrasting pictures may appear somewhat exaggerated, they reveal some of the multiple social benefits anticipated by advocates of rural electrification.

The critics of rural electrification consider the prospect of significant benefits to rural households unrealistic. According to them, rural electrification can benefit developing societies only through improvements in rural productivity and thus increase in rural incomes. Providing one or two light bulbs for a handful of probably high-income families through expensive rural electrification programs is impractical in the context of other development priorities. Even if some households benefit from electricity access, they question whether the benefits are worth the substantial investments and required subsidies. This chapter seeks a better understanding of rural electrification's social impact at the level of both the household and community. Among the many questions to be answered are whether rural electrification improves the quality of rural life, increases literacy and education and diminishes community outmigration.

The Issue of Causality

The India and Colombia studies were designed to gain some understanding of the complex relationship between rural electrification and household income. In those two countries, rural households are quite diverse. In Colombia, the study covered regions characterized by little irrigation, as well as a coffee-growing region with large plantations. For India, the study covered regions in the semi-arid tropics that feature extensive irrigation. The mean income and educational level in Colombia was much higher than in India. The breadth of population types provided sufficient variation to examine the social impacts of rural electrification.

The studies randomly selected households with and without electricity in occupations from high to low income to ensure a wide range of incomes in both groups. In households with electricity, total income was higher for every occupational category in India, and in Colombia, average income was higher in four out of five occupational categories (table 5.1). Thus, families with electricity had somewhat higher incomes on average, even within the same occupational categories. The same pattern was observed for education.

Because the sample contained relatively poor households with electricity and comparatively wealthy households without electricity, the research could sort out the impact of electrification by comparing households with essentially similar levels of income or education. As expected, households with higher incomes had higher levels of appliance ownership than poorer households. But, as will be seen in the next section, poorer households with electricity also had higher levels of both electric and nonelectric appliance ownership than households with similar levels of income without electricity.

Table 5.1 Income and Education by Income Class in India and Colombia, 1981

| | Income (US$) | | Education (years) | |
| | With | Without | With | Without |
Class by country	electric.	electric.	electric.	electric.
India				
Large farmers	2,712	1,618	5.6	4.9
Medium farmers	1,063	827	5.9	4.9
Small farmers	705	517	4.4	3.7
Shopkeepers/businesses	772	543	5.8	4.5
Artisans	644	379	3.1	3.0
Agricultural laborers	351	276	1.7	1.5
Average for all groups	1,161	645	4.8	3.6
Colombia				
Farmers	2,429	1,414	12.4	9.0
Shopkeepers/businesses	2,352	2,107	12.3	9.0
Government workers	2,826	3,042	12.2	7.3
Agricultural laborers	2,206	1,165	10.0	6.3
Average for all groups	2,563	1,497	12.0	7.5

Sources: India Survey 1981; Colombia Survey 1981.

Note: Income is annual level per household, and education is highest level achieved by adult family member.

Appliance Ownership

Appliance ownership can directly affect the quality of rural life. The use of such appliances as lights, stoves, televisions, and radios has a direct social impact, whether good or bad. Some consider television viewing good because it integrates households into the larger society, while others may criticize it for taking time away from productive activities or social interaction. Electric appliance use in rural households may be instrumental in changing rural energy-use patterns. Having better lighting may encourage households to cook with liquefied petroleum gas (LPG). Also, the

purchase of appliances might have secondary effects, such as stimulating development of a country's small- or medium-sized industries.
In the United States during the 1940s and 1950s, the tremendous
growth in demand for appliances in rural areas can be traced directly
to the presence of rural electrification programs. Rural electrification
was particularly important for catalog merchandising companies.
But in developing countries, an explosion of demand for appliances
may not occur because of their lower levels of rural household income. Thus, this section examines the use of appliances before and
after electrification for Colombia and India.

Household Appliances in India and Colombia

Appliance use is an important historical consequence of electrification and deserves explicit attention. Household appliances have
been cited as improving women's lives in rural households (Lay and
Hood 1979); that is, labor-saving appliances can decrease the drudgery involved in household work. For instance, one study found that
the four most often used electrical appliances for domestic work in
rural Costa Rica were electric irons, refrigerators, blenders and
washing machines (Goddard et al. 1981, p. 12). Of course, Costa Rica's per capita income is relatively high, and other less developed
countries may be unable to sustain the same level of appliance use.
For example, Tourkin, Weintraub, and Hartz (1981) found that only
a small percentage of households in Indonesia could afford appliances; however, two electric appliances—irons and televisions—were
quite common. Also, the purchase of rural appliances may have important indirect impacts through stimulating the development of
industries that produce consumer goods.

The similarities in appliance-ownership patterns for the India and
Colombia study samples were more striking than the differences (table 5.2). In both countries, appliance ownership was higher for
households with electricity, which can be partially explained by the
electricity-using nature of many appliances. As expected, the percentage of households owning appliances was generally higher for
Colombia, which during the 1980s had a higher per capita gross national product (GNP) than India. Because of Colombia's higher in-

comes, there were instances of higher appliance ownership in households both with and without electricity. But the surprising finding
was a substantially higher degree of appliance ownership than was
previously assumed for developing countries.

Table 5.2 Appliance Ownership in Rural Households in India and Colombia, 1981

Appliance ownership (% households)	India		Colombia	
	With electric.	Without electric.	With electric.	Without electric.
Lighting				
Electricity	100	--	97	1
Gas/kerosene	16	35	1	52
Candles	--	--	1	34
Flashlight	66	44	--	--
Stove				
Hot plate/electricity		0	23	0
Kerosene	39	12	28	16
Wood	---	--	25	56
Coal	--	--	IS	14
Leisure				
Electric radio	33	0	44	0
Transistor radio	47	39	42	65
Record player	3	0	--	--
Television	1	0	61	0
Households goods				
Iron	23	2	87	48
Electric iron	--	--	81	0
Other iron	--	--	6	47
Sewing machine	32	10	34	12
Table fan	32	.0	0	0
Ceiling fan	24	0	0	0
Refrigerator	1	0	30	2
Transport				
Bicycle	69	43	--	--
Motorcycle/moped	5	0	--	--

Sources: India Survey 1981; Colombia Survey 1981.

Note: (--) indicates data not available or not measured.

In Indonesia's Klaten region, the study also found higher-than-anticipated use of appliances beyond lighting. Nearly half of households with electricity had a non-lighting appliance, and one-quarter owned larger appliances, such as televisions, refrigerators, sewing machines and water pumps. This finding implies that electrification involved many appliances beyond just electric lighting, as had been previously thought.

The consensus of studies during the 1980s was that lighting was the primary benefit of rural electrification. According to several studies, electric lighting extended the day of households by two-to-three hours (Butler, Poe, and Tendler 1980; Mandel et al. 1981). The most commonly cited advantages of electric lighting compared to kerosene lamps was that they were brighter, less polluting and less expensive. Of course, cost comparisons depended on the level of subsidies for kerosene and electricity. Other consequences were an increase in household productivity and a greater feeling of safety and security for the family. In some instances, members of communities with electricity felt less isolated with the advent of electrification (USAID 1983).

Lighting was a necessity for households in both India and Colombia. Most households without electricity used kerosene lamps or candles during evening hours. After adopting electricity, the household immediately switched to electric lighting. But in India, some households with electricity still had kerosene lanterns, which were probably used as a backup lighting source in case of power failure. In India, only one in three households without electricity used kerosene lanterns, while the rest used candles or had no lighting at all. By contrast, in Colombia, half of households without electricity used kerosene and about one-third used candles. The more extensive use of commercial fuels for lighting in Colombia may have reflected the relatively higher household incomes.

Cooking accounted for a substantial percentage of energy use in these rural households. According to some estimates in the 1980s, cooking in rural households accounted for up to 90 percent of total household energy consumption, a pattern that continues even today. But rural areas of developing nations seldom use electricity for cooking. Fuelwood, charcoal, agricultural waste, dung and kerosene remain the main sources of cooking energy after electrification. In previous studies, only Costa Rica was identified as having made extensive use of electricity for cooking (Goddard et al. 1981), with about 22 percent of connected households cooking with electricity.

During the 1980s, traditional fuels were the principal cooking energy source in both India and Colombia; these included wood fuels, dung and agricultural waste for India and coal, wood fuels and charcoal for Colombia. But a substantial number of households with electricity in India used kerosene stoves because electric cooking would have required expenditures well beyond their means (Sen 1980). The same general patterns of traditional fuel and kerosene use held for Colombia, with the exception that about one in five households with electricity used electric hot plates or stoves, which was more extensive than anticipated. The households without electricity relied more on fuelwood and other traditional fuels.

The use of household appliances—irons, sewing machines, refrigerators and fans—was much higher in Colombia than in India. In India, about 1 in 4 households with electricity had an iron, compared to about 9 out of 10 in Colombia, reflecting that country's higher per capita income. Ownership of sewing machines was nearly identical for the two countries: about one-third of households with electricity and one-tenth without electricity owned sewing machines. Refrigerator ownership once again reflected the higher standard of living in Colombia, where just under one-third of households with electricity had refrigerators, compared to virtually none for India. Nearly half of India's households with electricity owned a table and/or ceiling fan, compared to only 1 in 10 for Colombia, characterized by a much cooler climate, especially in its mountainous regions. The extent of fan ownership in India was quite surprising, reflecting households' willingness to make a substantial investment in comfort during the hot summers.

Leisure time and communications are an important component of rural quality of life. Electricity made possible the greater use of plug-in radios, televisions, and record players. For India and Colombia, ownership patterns for either battery-powered or plug-in radios, televisions and record players were remarkably similar. For India, 33 percent of households with electricity had plug-in radios, as compared to 44 percent for Colombia. In both countries, households with electricity also owned a high number of battery-run transistor radios. Of course, households without electricity also owned transistor radios—about 1 in 3 in India, compared to 2 out of 3 in Colombia. For television ownership, the country trends differed significantly. During the 1980s, virtually no television broadcasting station sent signals to rural India, which explains why, unlike today, few Indian households with electricity owned television sets. In Colombia during the 1980s, by contrast, more than half of households with electricity owned televisions—a remarkably high percentage for what was then a relatively expensive luxury item. Today, nearly all households in Colombia, as in India, have televisions. As will be seen later, television ownership has had a significant impact on use of leisure time in Colombia.

Household Characteristics and Appliance Ownership

Households with electricity own more appliances because a whole new range of appliances becomes available for purchase. Television sets, fans, plug-in radios and electric irons simply are not possible without electricity. The question is, for households at the same level of income and education, would those with electricity own more appliances than those without?

Results of a previous analysis of appliance ownership in India and Colombia indicated that income, education and household electrification all make positive, independent contributions to household appliance ownership (Barnes 1988, p.102). This means that, in the 1980s, appliance ownership was higher for households with electricity than households without electricity, even for those with similar levels of income and education. Higher levels of household income, education and electrification all contributed to higher levels of appliance ownership, and consequently improved quality of life. The strength and similarity of the results from India and Colombia suggested that the relationships between income, education, electrification and appliance ownership were universal patterns.

The relationship between rural electrification and appliance ownership is also likely to grow over time. A study in rural Vietnam conducted for the same households between 2005 and 2007 found significant growth in appliance ownership over just three years (World Bank 2011). From 2002 to 2005, the number of color televisions and rice cookers owned grew by more than 25 percent. Black-and-white televisions, the only appliances to decline in ownership, resulted from upgrading to color sets. Ownership of water pumps and fans, both of which could have related health benefits for rural households, also grew substantially (table 5.3).

Table 5.3 Appliance Ownership for Households with Electricity in Vietnam, 2000–05

| Appliance | Households owning appliances (%) | | |
	2002	2005	Change
Electric fan	59	76	17
Color TV	48	74	26
Black-and-white TV	19	11	-8
Rice cooker	21	51	30
Water pump	21	32	11
Electric iron	20	28	8
Refrigerator	20	28	8
Computer	5	8	3

Source: Khandker, Barnes, and Samad 2013a.

To summarize, rural electrification makes a significant, independent contribution to appliance ownership. The level of appliance ownership in India, Colombia, and Indonesia during the 1980s was surprisingly higher than expected. Both Colombia and Indonesia saw high levels of television ownership, while India had a high level of fan ownership. Also, India and Colombia exhibited substitution of kerosene and candles with electricity for lighting. In terms of cooking appliances, some households with electricity in Colombia owned electric hot plates. For developing nations during the 1980s, electricity was generally considered too expensive to be an alternative energy source for cooking, a pattern that holds even today. In both India and Colombia, such household goods as irons and sewing machines were quite prevalent, but refrigerators were in use only at a modest level in Colombia. Appliance ownership generally was higher in Colombia than in India, as higher incomes facilitated greater appliance ownership.

Changes in Living Patterns

Advocates of rural electrification have indicated that major changes in activities and patterns of behavior occur in households with electricity. High-quality lighting provides more time in the evenings for study, working, social visits and entertainment, along with a host of possible productive activities. The most common finding in the literature on rural electrification is that, in households with electricity, children can read at night and there is a greater feeling of safety. In fact, the evidence already presented is quite persuasive in showing that, for lighting, electricity substitutes for kerosene or candles.

In India and Colombia, children's reading improved with rural electrification, but the findings are inconclusive on other important issues (table 5.4). The impact of electrification on the household behavior of men, women, and children appeared to have been more dramatic in Colombia than India, possibly due the effects of television ownership. In rural Colombia, both men and women greatly altered their living patterns to accommodate television viewing in the evening. In the study sample, nearly half of men and women viewed television in the evening, as did more than half of children. Thus, television viewing significantly altered the time devoted to other activities before household electrification. For instance, time spent reading declined substantially for adults, as did productive activities, social visits, and radio listening. Domestic work in the evenings also declined for women in the sample.

Table 5.4 Daily Activities Affected by Rural Electrification in India and Colombia, 1981

Household members, daily activity	India (time use)		Colombia (recall)	
	With electric.	Without electric.	After electric.	Before electric.
Men (minutes per day)				
Reading	--	--	18	35
Television	--	--	43	0
Social visits	34	38	29	68
Working	88	90	22	56
Radio	--	--	62	97
Women (minutes per day)				
Reading	--	--	16	47
Television	--	--	44	0
Social visits	18	15	35	70
Working	41	36	12	60
Radio	--	--	63	95
Domestic work	78	83	57	65
Fuel collection	9	19	--	--
Children (minutes per day)				
School attendance	38	36	--	--
Reading	42	33	72	43
Television	--	--	54	10
Working	9	6	0	0
Domestic work	15	15	23	46
Play	39	38	--	--

Sources: India Survey 1981; Colombia Survey 1981.

Note: For Colombia, figures are based on recall questions for households with electricity recollecting what life was like before electricity. For India, figures are from a time-use survey for households with and without electricity. (--) indicates data not available.

All of these activities were probably replaced by television view-ing. Obviously, television is a potentially powerful instrument for social change, but the study results in rural Colombia suggest that some of the consequences might be negative.

Even though television viewing seems to have caused a decline in reading among adults in rural Colombia, children substantially increased the time they spent reading in the evenings. According to the survey results, children apparently spent more time reading after households gained access to electricity; this was so despite the dramatic increase in the percentage of children watching television. In India, even without television sets, the time spent reading was substantially higher for children in households with electricity. The differences in the percentages of children reading sometime during the day or evening apparently was due to electrification of the household, which gave children more opportunity to read in the evenings. The decision to take a household connection might even have been swayed by the number of children in school.

The 1980s finding that rural electrification is related to children's increased study time has been reconfirmed many times by more recent studies. For instance, a 2002 national study on rural energy in Bangladesh (Asaduzzaman, Barnes, and Khandker 2009; Barakat et al. 2002) found that having electricity increased children's evening study time by about one-third of an hour and by more than half an hour during the day (table 5.5). This also was accompanied by increases in school attendance. No doubt, better prepared children were more likely to stay in school.

Table 5.5 Children's Study Time by Household Electrification Status in Rural Bangladesh, 2002

Children's study time (hours per day)	Household status	
	With electricity	Without electricity
Day	2.7	2.1
Evening	1.3	1.0

Source: World Bank 2008.

The pattern of electricity's relationship with rural time use was confirmed by later studies in India that also included television. Women living in households with electricity clearly had more balance between work and leisure time (World Bank 2004). Compared to women in households without electricity, they spent less time collecting fuel, fetching water and cooking, instead spending more time on earning income, reading and watching television.

The probability of women spending time reading was highly related to household electricity status, regardless of income class (table 5.6). The remarkable result was that women in households without electricity spent hardly any time reading, even in higher-income groups. As expected, the time spent reading for those who had electricity increased steadily with income. Apparently, having high-quality lighting at home is a virtual necessity for reading, from the lowest to the highest income groups. The general pattern was that women in higher-income groups take greater advantage of the benefits of rural electrification. However, even those in the lower-income groups led lives with less drudgery than those without electricity.

Table 5.6 Women's Reading Time by Household Electrification Status in Rural India, 1996

Household income class	Time spent reading (minutes per day)	
	Without electricity	With electricity
Very poor	3	6
Poor	2	9
Below middle	3	12
Above middle	3	16
High	1	19
Very high	1	19

Source: World Bank 2004.

Note: The time spent reading was for day of survey (11% of women responded that they had read that day). Most women who read did so for about 1 hour per day; results in the table are average for all women.

For the 1980s, differences in household time use were more dramatic in Colombia than India. But electricity still had an impact in India. Although social visits and working by women or men in India did not appear affected by electricity, the amount of domestic work done by women was lower for those with electricity. Fuel collection was substantially lower for households with electricity, suggesting that these households purchased, rather than collected, wood for cooking. More recently, it was found that households with electricity were more likely to use LPG (World Bank 2002b). But in the 1980s, households with electricity were more likely to use kerosene or purchase fuelwood for cooking than households without electricity, probably because electric lights would aid in cooking after dark. Cooking on a kerosene stove without kitchen lighting is more difficult than cooking over a fire, which provides much of the light for meal preparation.

Another difference between India and Colombia was that the Colombian households with electricity had changed their sleeping patterns (table 5.7). They tended to stay up longer at night, going to bed about 20 minutes later than they did before electrification. As discussed above, no doubt television was responsible for this change in household behavior.

Table 5.7 Waking Hours for Households with and without Electricity in India and Colombia, 1981

	India			Colombia		
	Hours		Minutes	Hours		Minutes
Household member group	With electric.	Without electric.	Diff.	With electric.	Without electric.	Diff.
Men	--	--	--	15.2	14.9	22
Women	15.9	15.8	-8	15.3	15.0	20
Children	14.0	14.3	-17	14.7	14.2	27

Sources: India Survey 1981; Colombia Survey 1981.
Note: (--) indicates data not available.

By comparison, India experienced no significant change in sleeping patterns. The comparisons between households with and without electricity indicated that men, women and children were going to bed at about the same time, approximately 9:30 p.m., and the length of their sleep was also the same.

To summarize, rural electrification seems to have had a substantial impact on social activities in rural Colombia. Adults tended to read less, not socialize as much and work less in the evenings. The social impact of television was obviously a two-edged sword. The negative effect was that evening social and productive activities were being replaced by television viewing. On the positive side, television viewing might have broadened the horizons of rural residents, introducing them to new technologies for work and new products through advertising. The introduction of any new technology is likely to have both positive and negative consequences for society; the debate over the relative merits of television has been a contentious issue ever since it became widely available in the 1950s. In India, with next-to-no rural television broadcasting, the results are not as dramatic. The positive impacts are that, in households with electricity, fuel collection occurs less, while time spent reading improves for children. The main reasons for the more pervasive impact of electrification on household behavior in Colombia versus India are probably linked to higher rural incomes and greater access to television.

Quality of Rural Life

The perception of the quality of rural life is quite a subjective notion summarizing impressions of many aspects of one's life. The previous discussions of appliance use and changes in living patterns should be reflected in quality-of-life impressions. Since rural electrification and income are associated with higher levels of appliance use, one might hypothesize that the perception of the quality of life should have been enhanced in those households with electricity and higher incomes.

In the India study, four survey questions were used to measure the perception of quality of life. Each respondent was asked to compare his quality of life to: (i) his father's time, (ii) five years ago, (iii) his neighbor's quality of life, and (iv) the quality of life he anticipated in the future (figure 5.1). Compared to their counterparts without electricity, the households with electricity appeared to have a perception of a generally higher quality of life.[9] The difference was fairly large for comparisons with their father's time, undoubtedly a period without electrification, and with their neighbor today, who probably would not have electricity. The conclusion was that the households with electricity perceive their quality of life as higher than households without electricity.

Figure 5.1 Quality of Life in Households with and without Electricity in India, 1981

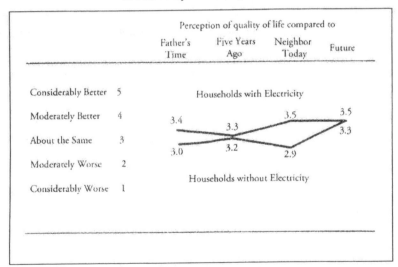

Source: India Survey 1981.

---

[9] Income probably also is related to the respondents' perception of the quality of their lives. A multiple regression was conducted, controlling for income and education. The results indicate that, even for households at similar levels of income and education, household electrification is still positively related to two measures of the quality of life (Barnes 1988, p. 108).

While electrification made a difference in the quality-of-life perception compared to one's father's time and one's neighbors, there was little difference in comparisons with five years ago and for the future. Of course, many of the surveyed households had had electricity for more than five years, which may explain the lack of significance for that time period. According to figure 5.1, the households with electricity perceived their present and future quality of life at about the same level. This contrasts with the households without electricity, who perceived that their future would be much better than their current situation. They may even have been anticipating the electrification of their home sometime in the future.

The conclusion from the quality-of-life questions was that electrification and income have important consequences for the perception of quality of life among these rural people, especially in comparison to the quality of life during the father's time and that of their neighbors. The households without electricity feel that there would be an improvement in their quality of life sometime in the future.

Migration

An important early justification of rural electrification was that it would lessen migration from rural to urban areas. Because of the "over-urbanization" problems experienced by some developing countries, reduction of rural-to-urban migration seemed quite desirable. The somewhat controversial concept of over-urbanization was that the growth of large metropolitan areas was occurring too fast in developing nations (Lipton 1980). The theory was that rural residents were being pushed toward the urban centers for lack of employment opportunities at home. The fear of over-urbanization stemmed from the growing number of slums and favelas in developing countries. The rural factors cited as causes of over-urbanization included all the social ills and difficulties of rural life, including lack of employment, poor infrastructure (e.g., schools and roads), little or no electrification and the general low quality of rural life in developing nations.

The corollary to the push theory was that urban areas acted like magnets pulling the rural populations to jobs, higher incomes and a better quality of life. Although all developed countries have undergone a substantial transformation from predominantly rural to urban populations, the theory of over-urbanization holds that the process was occurring much too fast in developing nations and that it would be prudent to stem the flow to enable a more orderly transition. Examples of urban congestion like that of Mexico City and Rio de Janeiro tended to reinforce the notion that stemming rural-to-urban migration would be desirable in developing countries.

Advocates of rural electrification expected that the anticipated improvement in the quality of life would keep rural populations from migrating and consequently alleviate problems associated with over-urbanization. The critics claimed there was no evidence to support the contention that rural electrification affected migration in any way. However, beyond anecdotal information, during the 1980s there had been no studies on the relationship between migration and electrification. Even today few studies have tackled the issue of the relationship between rural electrification and migration.

The assertion that migration from rural areas would be stemmed by rural electrification must be considered in the context of the myriad reasons for migration. First, most migration in developing countries is job-related. People typically migrate when the local economic opportunities are limited or when there are greater opportunities elsewhere. Second, seasonal migration must be distinguished from longer-term or permanent migration. Seasonal migration for harvests or other farm-related work is quite common in developing countries and typically involves rural-to-rural migration. Third, education can be a significant reason for migration from rural to urban areas since most educational opportunities are in towns or urban areas.

Rather than diminishing migration, one possibility was that rural electrification would increase outmigration. At the time rural areas in industrial societies were receiving electricity, massive rural-to-urban migration was taking place. Thus, the expectation that rural electrification would cause socioeconomic development in rural areas seems inconsistent with the assumption that it would reduce rural outmigration. In the United States, for example, migration from rural areas occurred overwhelmingly at the time of its rural electrification programs. Electrification often raised rural expectations, and young people sought wider opportunities in the form of jobs or education in urban areas. As a society's level of specialization increases, one would expect a general increase in migration. Thus, the historic pattern of increased rural-urban migration in industrialized societies was contrary to the accepted wisdom of rural electrification advocates.

Migration in India

During the 1960s, India experienced a fairly small shift of rural populations to urban regions. In both rural and urban areas, population increased at about the same rate. Urban industries grew steadily, while agricultural growth was tremendous. The modernization of agriculture did not displace jobs; rather, it expanded them, sometimes even causing job shortages at harvest time. Thus, employment opportunities were created for laborers due to agricultural development; but the wealthier farm families may have lost members because of migration for education and more specialized jobs in cities.

The purposes for which persons migrated included such activities as education, seasonal labor and government jobs (table 5.8). More than two-thirds of permanent migration was for government employment or other reasons. By contrast, only a small percentage of permanent migration was for work in agriculture. About one-quarter of the migration was for obtaining a higher education, so it was a very important reason for moving from a village.

Migration for education was significantly lower in villages with electricity. Of course, villages with electricity might have had better educational facilities or teachers willing to stay in a village, possibly lowering the number of migrants from such villages. By contrast, migration for government service was significantly higher for villages with electricity, partially confirming that electrification might increase, rather than decrease, migration.

The findings of the 1980s surveys were contrary to the expectation of advocates of rural electrification, but were in line with the reasoning that rural electrification should raise the expectations of villagers. The villages with electricity had more permanent outmigration, but less seasonal outmigration for work than the villages without electricity. Permanent outmigration typically involved movement to towns or cities. The percentage of total permanent migrants moving to towns was 88 percent for villages with electricity compared to 78 percent for villages without electricity.

Table 5.8 Percent of Population Migrating in Villages with and without Electricity in India, 1981

Population group (%)	Villages in India	
	With electricity	Without electricity
Village population that migrates for:		
Permanent work	2.39	1.23
Seasonal work	2.28	3.93
Permanent migrants moving to:		
Town	88.62	78.75
Village	11.37	21.24
Permanent migrants moving for:		
Education	20.97	29.24
Agriculture	4.86	4.07
Government service	44.09	34.53
Other reasons	29.55	30.29
Permanent migrants with education of:		
Below 5 years	26.17	18.58
Above 5 years	73.83	81.42

Source: India Survey 1981.

Seasonal migration was typically rural-to-rural migration and involved migration for harvests. Seasonal migration was lower in the villages with electricity, probably because of an increased demand for labor due to more extensive irrigation and higher crop yields. This seems to confirm that farming land for more than one crop during a year improves local occupational opportunities and reduces village outmigration for seasonal labor. Although seasonal migration was somewhat related to electrification, the conclusion was that other factors, such as percent of double-cropping in areas of the villages, were probably more important.

Permanent migration in villages with electricity actually increased. The main reason for permanent migration was for employment, especially in government service. The notion that permanent migration would be deterred by rural electrification could not be confirmed by the India research. On the other hand, the increase in outmigration for communities with electricity should be considered consistent with the goal of economic development. If electrification stimulates small-scale industrial and agricultural development, then migration both in and out of rural communities would be greater.

Migration in Colombia

Migration in Colombia also was predominantly related to employment opportunities. Santamaria (1980) found that Colombians tended to move for economic reasons; nearly two-thirds of the immigrants in that study had left their birthplace to seek higher wages, better jobs or the possibility of owning land. The commercialization of agriculture in Colombia, especially plantations in the coffee-growing region, resulted in an expansion of the labor force. This caused quite distinct regional-population movements, with persons migrating from the central areas to the coffee regions. Migrants were moving to regions with a higher demand for skilled and unskilled labor and were settling down in small towns. Leading up to the 1980s, the number of small cities with populations of 20,000–30,000 had tripled.

The length of residence should reflect the degree to which newcomers move into the community since those with quite stable populations would have few immigrants. In Colombia, the families in households with electricity had not resided in the community as long as those in households without electricity. More than 13 percent of the households with electricity had arrived in the community during the previous 5 years, compared to 8 percent of the households without electricity (table 5.9). Similarly, the proportion of residents who had lived their entire lives in the community was 49 percent for households with electricity and 55 percent for those without, meaning that communities with electricity tended to have more dynamic populations, though the differences were somewhat small.

Table 5.9 Migration for Communities with and without Electricity in Colombia, 1981

| | Communities | |
	With electricity	Without electricity
Household migration type		
To community (%)		
Arrived in last 5 years	13.8	8.4
Arrived more than 5 years ago	36.9	36.7
Living entire life in community	49.0	54.9
From community (%)		
With one member who left	35.1	29.9
Reason family member left:		
Obtain education	2.3	6.5
Work opportunity	61.7	51.6
Family reasons	16.3	22.6
Other reasons	19.5	19.4

Source: Colombia Survey 1981.

Community outmigration was more difficult to measure as families that had moved out of the communities could not be interviewed. However, respondents were asked whether "a household member had left the community" and the "reason for migration." Communities with electricity were more likely to have lost population than those without electricity. The reasons stated for family members' having left the community were overwhelmingly related to jobs and income: 61 percent of households with electricity and 51 percent of those without cited economic reasons for community members' migration. By contrast, family reasons accounted for about one-fifth of total outmigration, and the number migrating for education was low. Communities with electricity tended to have both a slightly higher influx and a higher outflow of population. The caveat is that migrants in Colombia were moving into the predominantly electrified coffee-growing regions, but movement within those regions was unrelated to electrification. If rural electrification had an impact on migration, it would have to be indirect through the creation of new jobs, essentially in coffee-growing areas.

The conclusion of the 1980s research was that the most important factor in migration decisions in both India and Colombia was employment opportunities. Over 60 percent rural households in India and Colombia reported that the primary reason for migration was employment or income-related activities. In India, electrification resulted in outmigration to towns and cities to obtain jobs and higher levels of education. There also was less seasonal outmigration, probably because of the improvement in local agricultural employment. In Colombia, employment opportunities in the extensively electrified coffee-growing regions caused an influx of new residents to the communities, but this was apparently offset by the lack of migration within the regions. In India, village electrification increased migration where the electrification programs had had productive impacts. In Colombia, where the productive impact of electrification had not materialized, electrification of communities had not caused dynamic movement of rural populations. Thus, the effect of electricity on migration may have had more to do with its impact on local employment as compared to making life better in rural communities.

Education and Literacy

The role of education and literacy in rural development has been an important and somewhat controversial issue in the development literature. Many consider education to be a prime mover of development (Lerner 1958; Schultz 1964). The consensus is that literacy and education are fundamental for a society's advancement to higher levels of development and a more complex division of labor. However, some had questioned whether literacy is the cause or the effect of development (Barnes, Fliegel, and Vanneman 1982). Regardless of the direction of causality, no one would question that education is one of the key "human-capital" inputs for societies at higher levels of development.

A common finding among rural electrification studies was that rural electrification, income and education were all positively related (Madigan 1981; Madigan, Herrin, and Mulcahy 1976; Herrin 1979; Saunders et al. 1978). These studies implied that electrification was one of the causes of higher income and educational levels. However, an alternative explanation might have been that wealthy households who were better able to afford electricity probably adopted the service at a higher rate than poorer households. In this study, it was possible to control for income in the Colombia household analysis and for village level of agricultural development in the India study. Therefore, the relationship of rural electrification, level of development and education can be disentangled.

Electric lighting is attractive for educated adults and families with school-aged children because it allows for reading during evening hours. In both India and Colombia, the study findings showed that rural electrification led to higher levels of children reading at night. In addition, rural schools and teachers may have been more likely to locate in communities with electricity, thereby providing the infrastructure for increasing rural literacy over the long term. Thus, if rural electrification stimulates the growth of education, and if highly educated adults are more likely to adopt an electricity connection, then rural electrification may well be both a cause and an effect of rural literacy.

In India and Colombia, households or communities with electricity had higher levels of literacy and education. In India, the average proportion of literacy in villages without electricity was 19 percent (table 5.10). The figure was slightly lower for newly electrified villages, but improved for villages that had had electricity for longer. The number of children attending school as a percentage of total school-aged children no doubt had long-term consequences for the rural literacy rate. While village leaders may have exaggerated the number of children attending school, school attendance mirrored the literacy rates from the census materials. Among households that had had electricity for less than 5 years or were without electricity, less than half of eligible children attended school, while well over half of eligible children attended school in villages that had had electricity 6 or more years.

The relationship between rural electrification and education was quite similar for India and Colombia, but the level of education in Colombia was substantially higher than in India. However, as in India, the level of education for the Colombia sample improved significantly with years of electrification. In communities with 19 years of electrification or more, household heads averaged 5 years more schooling than those in communities without electricity. The number of newspapers and magazines read by the family was higher in communities with electricity.

Table 5.10 Rural Electrification and Indicators of Education in India and Colombia, 1981

	India			Colombia	
Electricity in village	% literate in village	% eligible children attending school	School proximity index	Household head's years of education	No. of newspaper magazines per household
No electricity	19	48	3.4	8.1	1.5
1–5 years	17	46	4.5	9.4	1.8
6–10 years	21	54	5.0	12.2	2.2
11–15 years	26	67	5.1	13.3	2.2
16–20 years	23	59	4.8	--	--
20 and above	26	52	6.5	--	--
Correlation	0.25	0.17	0.39	0.31	0.21

Sources: India Survey 1981; Colombia Survey 1981.
Note: (--) indicates not applicable.

Since the 1980s study, the impact of rural electrification has been the subject of numerous subsequent studies, and most of the Colombia and India findings have been verified. For both India and Colombia, rural electrification and education programs had mutually reinforcing consequences. Whether by providing electric lights for reading or improving the environment of village schools, electrification and education were associated, even when taking into account household income and village level of development. In more rigorous recent studies, rural electrification has been found to have a significant impact on school enrollment and children's study time (chapter 9). In Bangladesh, the increase in children's study time due to electrification was more than 20 percent for boys and 12 percent for girls (Khandker, Barnes, and Samad 2013b). Likewise, in Nepal, the increase in study time resulting from having electricity from micro-hydro systems was 7 percent for boys and 12 percent for girls (Banerjee, Singh, and Samad 2011). In both India and Vietnam, school attendance rose more than 6 percent for both boys and girls due to household electricity (Khandker, Barnes, and Samad 2013a). The conclusion is that rural electrification and education are complementary programs. Schools are not as effective in educating children well without electricity and electrification would be unable to enhance the education of rural children without local schools.

Growth of Household Connections

The anticipated rate of growth in household connections provides planners with a basis for assessing the financial viability of rural electrification programs. During the 1980s, the connection rates for developing countries were somewhat disappointing (PEO 1965), bringing into question whether rural electrification projects were economically justified. Thus, understanding the factors that contribute to faster-than-average growth of household electricity adoption is important to the success of rural electrification programs. Electricity company data for household electricity in Colombia and Indonesia proved to be unreliable, so the following discussion is confined to India.

For the 90 villages in the India 1980s survey that already had electricity, the growth pattern of household adoption was examined for the previous 20 years. The villages experienced an initial surge in electricity adoption by rural households for the first three years after a village received electricity, probably reflecting pent-up demand (figure 5.2). For each of the first two years of electrification, the villages experienced an adoption rate of 10 and then 7 households per year. Afterwards, the household adoption rate steadied at about 4 households per year. This means that, for an average village, the number of households with electricity increased from about 10 in the first year of electrification to more than 86 in the nineteenth year. Thus, the cumulative household adoption of electricity continued over a long period.

Figure 5.2 Cumulative Growth of Household Electricity Connections in India, 1981

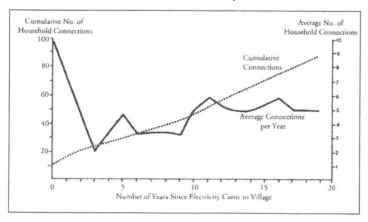

Source: India Survey 1981.

As expected, the passage of time and high levels of village populations brought about a long-term increase in electricity adoption. Literate households seemed to have had an added incentive to adopt electricity, simply because they could take advantage of electric lights for reading, underscoring the notion that rural electrification and education are mutually reinforcing programs. The idea that household adoption stops after a few years was found to be mistaken. The slow continuous adoption of electricity in rural areas continues until it reaches a saturation point.

Conclusion

In the 1980s—and even among some today—there was an impression that rural electrification implied only electric lighting for a small percentage of rural households. That impression remains wide off the mark. For India, Colombia and Indonesia, the positive benefits of rural electrification included diverse appliance use, more reading especially among children, and a higher quality of life; for India, it also included decreased seasonal migration. More than 60 percent of households with electricity in Colombia had televisions, and more than 40 percent of families with electricity in India owned fans. Perhaps because of television ownership, the living patterns in Colombia changed substantially due to electricity. The lifestyle changes for India included children's reading more at night, a subjective perception of a higher quality of life in households, and improved levels of education.

Rural electrification, in some instances, fell short of expectations. The anticipated benefits for productive work at night and more social visits did not materialize in Colombia. The significant increase in television viewing in Colombia may have been responsible for curtailing productive activities and reading among adults. In India, electrification did not alter sleeping hours or change work patterns during the day. In Colombia, migration was only marginally affected by electrification, while in India it may have contributed to some permanent village outmigration for work or government service. In Colombia, some concern was raised about the cost of electricity connections and television sets, all highly desired by rural households, taking money away from more basic needs, such as education and food for poor households.

The overall conclusion was that the social benefits of electricity are quite substantial. The evidence suggests that electricity had an impact on rural households, even after taking level of income into consideration. Rural households were willing to sacrifice a substantial amount of income to acquire electricity. The criticism that electricity had only superficial social impacts for rural households was quite wide off the mark. However, another issue remained. The benefits of rural electrification programs might not have been equally distributed among rural populations. The equity effect of rural electrification is the topic of the next chapter.

6. Equity and Poverty

The Achilles heel of many rural development projects is that their social and economic benefits are unequally distributed. Large irrigation canals often benefit mainly large farmers; influential community members monopolize credit programs. Agricultural modernization programs can contribute to land concentration, pushing small farmers into the ranks of the agricultural laborers. Rural development extension officers tend to concentrate their efforts on influential or "innovative" farmers. Many development agencies have adopted policies to help the "poorest of the poor." The question is whether the "poorest of the poor" can benefit from large, capital-intensive projects such as rural electrification. Does rural electrification exacerbate or ameliorate such rural inequality in developing nations? These questions and their answers are just as valid today as they were 25 years ago.

The advocates and the critics of rural electrification are nowhere further apart than on the question of rural inequality. Advocates contend that rural electrification is an instrument to redress inequalities in developing nations, as the contrast between rural and urban amenities is due, in part, to the availability of electricity. Frank Denton (1979, p. 2) graphically depicted the differences between urban areas with electricity and rural areas without electricity in the Philippines. He observed that the cities have department stores, good hospitals, neon lights, nightclubs, movies and factories. By contrast, in the countryside where there is no electricity, all activities cease at sunset because of the poor alternatives to electric lights, so rural families have unproductive and unentertaining evenings. Advocates reason that the stark urban-rural differences in developing nations are a consequence of poor infrastructure and low productivity and that electricity can improve the rural standard of living, moving it closer to those in urban areas. They further assert that delivering electricity to rural households, farms and businesses will not only close the gap between urban and rural lifestyles; it can also narrow the class distance between rich and poor in rural areas by creating new jobs and permitting the rural poor to raise their economic status.

The critics object, saying that the expensive electricity distribution systems will serve only the wealthiest families and thus reinforce existing inequities and wealth distribution. The adoption of electricity by rural households is highly dependent on income level. Thus, while the rich will be able to partake of all the benefits of electrification, such as the use of modern appliances and lighting, the rural poor may not be able to afford electricity. In fact, they may not even be permitted to have access to electricity because their houses are of substandard quality. In this context, Smith (1980, p. 87) argued that rural electrification goes to the wealthier regions, to the wealthier villages in these regions and to the wealthy households within these villages. In addition, critics claim that rural electrification has no impact on rural productivity and that the money allocated to the construction, generation and maintenance of expensive rural electrification systems would be better spent on projects that more directly improve the lives of the rural poor.

A new line of criticism today comes from those advocating renewable energy for rural electrification. Instead of providing electricity from national grids, they claim it is better to generate and distribute electricity from decentralized systems controlled by households or communities. But these systems are expensive; often only the richest communities or households can afford them, so they come under the same criticism as grid rural electrification.

This chapter examines the equity impact of rural electrification from various perspectives. The first is whether there was a relative widening of class differences in the 1980s studies. The second is whether land was being concentrated into fewer farms because of the greater ability of large farmers to take advantage of electric power. Third, electricity may have had differential impacts on women and children. Finally, electricity might have had a greater impact in developed rather than less developed regions.

Equity within Rural Communities

The cause-and-effect relationship between rural development, poverty and income distribution may take several forms. One possible scenario is that development improves real income while narrowing the gap between rich and poor, which is reported to have occurred in Taiwan during the 1960s (Chinn 1979). Conversely, real income might actually decrease for the rural poor during development, which apparently happened in Java during the 1960s (King and Weldon 1977). The most common pattern for developing countries is for both the poor and the rich to experience increases in real income, but with an increasing gap between socioeconomic classes.

Existing evidence suggests that the direct impact of rural electrification for rural households, especially over the short term, worsens rural inequality. While the poor are not totally excluded, in nearly all countries, the poor adopt electricity at a lower rate than wealthier households. For instance, in the Philippines, it was estimated that households with incomes below the poverty level could not afford electricity (Mandel et al. 1981). However, a Philippines survey at about the same time found that, while households with incomes above the poverty line adopted electricity at a higher rate, about one out of five households below this level did adopt it (Madigan 1981, p. 11). In higher-income countries, such as Costa Rica, the adoption rate is even greater. About seven out of ten households within reach of the lines adopted electricity, and more than half of these households were below the poverty line (Goddard et al. 1981). The household adoption rate for India was much lower, reflecting lower incomes, but the rate still improved as incomes rose. Thus, it is evident that income had a very important impact on whether a family adopted electricity. This was true both within and between countries; poor countries, like poor households, had lower adoption rates.

This strengthens the general argument against rural electrification—that it would improve the condition of wealthy families but have little impact on the rural poor. To better assess the equity impact of rural electrification, this section examines who adopted electricity and how it affected their quality of life, whether electrification adversely affected land equality, and the nature of its long-term relationship with rural poverty.

Income Class and Electricity Adoption

Generally, rural electrification programs in communities with high electricity-adoption rates are more equitable. Lower-income families adopt electricity at a much lower rate than wealthier households. The factors discouraging the adoption of electricity by the poor include high connection charges, high electricity prices and poor housing conditions. Yet for Colombia, household electrification was so pervasive, especially in the coffee-growing region, that the electrification programs benefited both rich and poor households (table 6.1). Because household income in India was much lower, so was the adoption rate.

Table 6.1 Regional Electricity Adoption Rates in India and Colombia, 1980

Country, state or region	% with electricity
India	
Punjab	56
Andhra Pradesh	14
Maharashtra	11
West Bengal	4
Colombia	
Coffee	88
Coastal	58
Central	42

Sources: India Survey 1981; Colombia Survey 1981.

This distribution of rural income also appears to have had an impact on the extent to which communities adopted electricity. The existence of a large middle class, as opposed to extremes of poverty and wealth, would be a much more favorable climate for household adoption. A wealthy region with a high degree of inequality would sustain only very low levels of electricity adoption. Results from the India survey demonstrated this point. During the 1980s, the relatively wealthy Punjab region had the highest rates of electricity adoption, while such poorer states as West Bengal and Andhra Pradesh had lower levels. The Punjab typically had a more equal distribution of income than did Andhra Pradesh. The low rate of electrification in West Bengal may well have been due to the high degree of regional inequality, but the state's electrification program had other problems, including extremely low reliability, which also discouraged electricity adoption.

Lower-income households may also have been unable to mobilize the necessary financial resources or lacked the influence to obtain power from the electricity authorities. For persons in India's lower occupational categories, it could take up to 10 years to obtain power supply, compared to an average of 6 years for those in the higher-income occupations (Samanta and Sundaram 1983, p. 202). This occurred despite the survey respondents' assertions that electricity went to "anyone who wanted it" rather than "mainly to influential persons" in the village. In Indonesia, the perception of most households was that wealthy households benefited more than poor ones (Brodman 1982, p. 71), implying that it was harder for the poor to afford or obtain electricity for their houses. Thus, it is apparent that the benefits of electricity went primarily to wealthier households in the communities, at least over the short term.

Consequence of Electricity Adoption

After adopting electricity, wealthy households also may derive more benefit from electrification since they are able to purchase more appliances and thus improve their quality of life. Of course, this is not unique to developing societies. Whereas lower-income families in developed societies may be unable to afford such items as

central air conditioning, lower-income families in developing societies similarly may be unable to purchase such items as radios and
fans.

Households in countries with high incomes can afford to purchase more appliances. Appliance ownership depends on household
income and, to some extent, culturally defined consumer behavior.
For instance, irons were valued in both India and Colombia, but the
ownership of irons was higher in Colombia (tables 6.2 and 6.3).

**Table 6.2 Appliance Ownership and Income Class in Colombia,
1980**

% owning	Rural income class		
appliance	Low	Medium	High
Iron			
With electricity	72	86	93
Without electricity	43	44	46
Sewing machine			
With electricity	21	23	41
Without electricity	16	11	20
Radio:			
With electricity	81	90	84
Without electricity	58	48	62
Television			
With electricity	39	53	39
Without electricity	0	3	0

Source: Colombia Survey 1981.

Within both Colombia and India, the ownership of irons was
highly dependent on the level of household income. A greater percentage of high-income families owned irons compared to low income families in both countries. Because Colombia generally had
high levels of income, even the lowest-income households were able
to afford irons compared to those in India. On the other hand, sewing machines were highly valued in India. Consequently, even
though household incomes were lower, the percentage of households

with sewing machines was about the same for both countries. But once again, high-income classes were more likely to own sewing machines.

Table 6.3 Appliance Ownership and Income Class in India, 1980

% owning appliance	Rural social class		
	Agricultural laborer	Small farmer	Large farmer
Iron			
With electricity	19	31	45
Without electricity	0	18	15
Sewing machine			
With electricity	19	24	42
Without electricity	7	6	10
Radio:			
With electricity	42	56	82
Without electricity	22	36	62
Television			
With electricity	0	0	0
Without electricity	0	2	0

Source: India Survey 1981.

Electrification also may play a role in increasing the variation in the number and types of appliances owned. Households without electricity do not have as wide a range or choice of appliances. For instance, in Colombia, ownership of electric irons varied with income for households with electricity, but households without electricity at all income levels owned few nonelectric irons. This suggests that electrification may have stimulated the introduction of new consumer products, whose purchase was related to the level or household income. In Colombia, television ownership was related to income, but ownership was not possible without some form of electricity. Also, more recent evidence suggests that households with televisions often allow their neighbors without televisions to drop by and watch various shows with them (World Bank 2002a). This line of reasoning suggests that, in regions with electricity, both rich and poor households are prospective purchasers of new appliances. The distribution of electric appliances favors the rich, but *the level of benefits for the rich and the poor households with electricity is higher.* This scenario shows appliance ownership improving for all classes, but with a larger gap evolving between rich and poor.

Lighting, irons, radios, fans and other appliances all can have an impact on people's perceptions of the quality of rural life. For the India sample without electricity, quality of life was perceived as much higher by large farmers, compared to the other occupational groups (figure 6.1). Low-income agricultural laborers without electricity rated their quality of life the lowest.

Figure 6.1 Quality-of-Life Perception of Households without Electricity in India, 1981

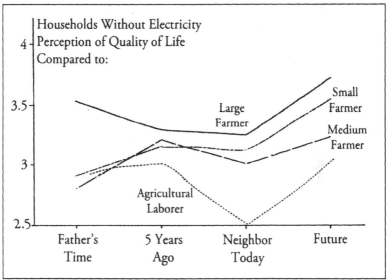

Source: India Survey 1981.

Electrification appears to have increased the overall quality of life and to have narrowed the gap between middle- and upper-income households. The perception of quality of life for those with electricity was both higher and more heterogeneous than among the sample without electricity (figure 6.2). The large farmer with electricity still rated quality of life higher than other groups for the past, present, and future; but the assessments of those in other occupations with electricity were closer to that of the large farmer, compared to the assessments of those in the same occupations without electricity. Thus, middle-income households with electricity had narrowed the gap on large farmers concerning quality of life. However, the lives of small farmers and agricultural laborers were still fraught with much uncertainty. While quality of rural life appears to have improved somewhat for the poor, it still lagged behind that of middle- and upper-income groups.

Figure 6.2 Quality-of-Life Perception of Households with Electricity in India, 1981

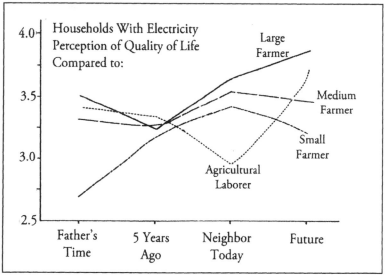

Source: India Survey 1981.

These lowest-income groups were at the bottom in their assessment of quality of life. The small farmer felt that the past was much worse than the present, perhaps reflecting upward mobility; but they also were fearful of the future. The agricultural laborer, by contrast, perceived that the past was better than the present, but was optimistic about the future. To summarize, with rural electrification, quality of life for middle-income households appears to have improved both relatively and absolutely. The poorest households with electricity rated their present quality of life low, but were optimistic about the future.

Land Inequality

During the 1980s, the transformation from traditional to more modern forms of agriculture raised the specter of capitalist agriculturalists buying out smaller farmers and displacing tenants, resulting in greater land concentration. Many documented historical cases of agricultural change reinforce this notion. The English enclosure movement was responsible for transforming peasants into urban laborers in England during the 19th century. During the same period in Japan, despite policies to encourage farmers to stay on their farms, rural landlessness increased due to commercialization of agriculture and population growth.

Rural electrification conceivably could cause increasing landlessness and inequality through the modernization of agriculture. Electric pumps and threshers could displace labor and enhance the income of large farmers. Because of this improved income, large farmers might then purchase new land from marginal farmers who cannot afford the capital expenditures to improve their farming practices. As indicated, the small farmers and landless laborers are in a very vulnerable position within Indian society. The result would be an increase in land concentration among large farmers with electricity.

If this scenario is true, then the number of years that a village has electricity should be related to higher village land inequality. But India has land ceiling and reform laws that might prevent this. Although the land reforms were rather ineffective in land redistribution, they did prevent many large farmers from acquiring new land (Ladejinsky 1973, p. 542).

The evidence from India suggests that rural electrification was unrelated to land inequality. Village leaders were asked to identify the ten largest farmers and estimate how much land they owned in order to construct a measure of land concentration. If rural electrification were related to land inequality, then year of electrification would have been positively associated with land concentration. Neither rural electrification year nor agricultural development adversely affected land distribution in India.[10] This may have been due to the land ceiling laws, which had been ineffective in preventing large farmers from retaining land, but prevented many from acquiring new land. Other studies also confirmed that agricultural development was not a cause of greater rural inequality in India (Barnes and Vanneman 1983; Attwood 1979).

Colombia is a country with large regional differences in size of landholdings. The coastal region is characterized by modern agriculture, cattle farming and large landholdings. The coffee-growing region is oriented toward export agriculture, but has much smaller farms than the coastal region. The central region is distinguished by traditional farms and small landholdings. One might expect the Colombia farm-level analysis of electrification and size of landholding to reveal that, during the 1980s, large farms were among the first to adopt electricity. But neither education nor year of electrification was associated with greater-than-average farm size. Rural electrification in Colombia was not found to have had much impact on agricultural productivity due to the lack of irrigation and thus perhaps would not be expected to play a significant role in increasing farm size.

Land inequality during the 1980s was neither exacerbated nor apparently improved by rural electrification. In fact, some contrary evidence suggests that land fragmentation in India was alleviated by investments in irrigation facilities (Samanta and Sundaram 1983). Irrigation encouraged farmers to own adjacent plots of land. To conclude, factors other than rural electrification appear to have been responsible for land inequality, as evidenced by the differing patterns of land ownership in the states of Colombia and reduction in land fragmentation in India.

---

[10] The results for India and Colombia in this section are based on analysis presented in the first edition (Barnes 1988).

Rural Poverty in India

Rural poverty in India involves such issues as agricultural productivity, nonfarm employment opportunities, minority groups and landholdings distribution. The classic study of rural poverty in India by Dandekar and Rath (1971) identified land distribution and employment opportunities as the key factors related to rural poverty. Their reasoning was that the means of production in rural India was ownership of land. Therefore, an equitable distribution of land among cultivators, along with high agricultural productivity, would result in less village-level poverty in several ways. High agricultural production would benefit landless laborers (scheduled castes and other minorities) since farmers would hire labor to perform tasks previously done by family members. In the process, the farmers would be required to pay higher daily wages to farm workers because of the increasing demand for agricultural labor.

Several crude indicators of rural poverty were available from the India survey, including the percentage of village population paying taxes on their property and the percentage of village population that leaders identified as "living in acute poverty." Villages that had electricity the longest also had less poverty than those without electricity, perhaps due to higher agricultural yields. On the other hand, rice-growing areas had a higher percentage of people living in poverty. Traditionally, rice-growing regions had a poor record of income distribution. In any event, rural electrification did not appear to be associated with rural poverty.

Women and Children

Existing evidence also indicates that women and children benefit most from household electrification. As was indicated in the previous chapter, appliances may lessen the work of rural women, while electric lights facilitate children doing school work at home. Supplementing this evidence, two studies in Costa Rica found that rural women benefited especially from electricity (Lay and Hood 1979; Kessler et al. 1981). Of course, Costa Rica is a high-income country with a high rate of electricity adoption. In less developed nations, women and children living in poor rural households unable to afford electricity would not directly benefit from electrification.

Some studies have found that agricultural development has had an adverse impact on rural women (Tinker 1976; Boserup 1970; Barnes 1983), but women and children apparently have been the prime beneficiaries of rural electrification. Lights and appliances can have a significant impact on household work. In the United States, subsidies were justified in the early stages of rural electrification, in part, because of improvements in the quality of rural life, including the elimination of household drudgery through the use of electric appliances. In 2002, after interviewing Louisan E. Mamer, one of the first organizers of efforts to promote machinery and appliance use in rural areas, Gina Troppa observed:

> The REA (U.S. Rural Electrification Administration) Farm Show was often referred to as the REA "circus" because it was usually held outdoors in large tents. The show was a popular and effective means of promoting electric cooperative growth, farm and home electrical equipment, and the wise use of electricity. Farmers and their families attended the events by the thousands. All were eager to learn more about the modern labor saving benefits of electricity and electrical appliances. 'One of the reasons for starting the farm show was that it was difficult to get electrical farm equipment to show in rural areas. So, we got it from the manufacturer and carried it on a truck from one place to another,' says Mamer. The REA circus and Mamer's demonstrations were effective. Some farmers would purchase products on the spot because they were so amazed at how easily the equipment worked. The first thing people wanted installed in their homes was ceiling lighting, but women were also interested in irons, washing machines and refrigerators. Mamer recognized the impact that electricity would have on women in the home, and her efforts greatly affected rural women's lives. (Gina Troppa, "The REA Lady: A Shining Example," *Illinois Country Living* [2002], paraphrased)

Even though appliance ownership in developing countries is much lower, rural electrification programs, especially those with household emphasis, still may have substantial benefits for women and children. A study in India during the late 1990s that measured women's time use in hours per day found that the drudgery involved in domestic chores declined for those women in households with electricity (table 6.4). They spent less time collecting fuel, fetching water and cooking. The mechanisms for this change might have been related to having better lighting in the evening. The households without electricity tended to prepare meals during the afternoon when light was plentiful, and then warmed the food again in the evening, perhaps causing them to spend more time cooking and collecting fuel.

Table 6.4 Women's Domestic Work for Households with and without Electricity in India, 1996

| | Communities | |
| | With electricity | Without electricity |
Women's time use (hours per day)		
Collect fuel	0.5	0.9
Fetch water	0.9	1.0
Cook	2.6	2.9

Source: World Bank 2004.

For rural communities in Colombia, electrification meant the freeing up of time for more recreational activities, with an emphasis on television watching. After electrification, family members stayed at home more. For teenagers and children, the most significant benefit was that more of them spent some evening hours studying. Women did evening chores during the day, and some domestic work was lightened through use of appliances. In homes without electricity, women anticipated doing more productive work in the evenings; but once they adopted electricity, these positive expectations were not fulfilled. Women worked less in the evening and spent more time on leisure activities. Since women and children spent more time in the home compared to men, it is quite reasonable that they should have benefited the most from household electricity.

To summarize, on the question of household equity from rural electrification programs in the 1980s and even today, the news is both good and bad. The bad news is that, in countries with extremely low incomes or poor records of income distribution, the poor were unable to afford electricity. Even among households with electricity, the wealthier were able to purchase more appliances, thus widening the gap between rich and poor. The good news is that, for those households that adopted electricity, their overall quality of life was enhanced compared to households without electricity, and, to some extent, the gap between middle-income and wealthy households was narrowed. Women and children, who spent the most time in the home, benefited more from rural electrification than did men, which was not the case in most rural development programs not specifically directed toward women.

Regional Inequality for Indian Agricultural Productivity

Politicians have long recognized that development projects can benefit their constituents, and rural electrification projects are no exception. Since large infrastructure projects typically generate additional employment, the direct benefits of rural electrification to regions would include the creation of new jobs. Also, rural electrification was thought to improve productivity in agriculture and industry.

Rural electrification might exacerbate regional imbalances by helping developed areas to improve faster than less developed ones. However, in the early 1980s, there was little empirical evidence to support or reject this contention. One study in the Philippines found that the advanced areas appeared to achieve much more economic development than the poor ones (Mandel et al. 1981). Unfortunately, this study examined advanced and less-developed areas without analyzing trends over time, so it was impossible to determine the differential impact on the two areas. While it may have appeared that electrification had little impact on the poor regions, in reality, they may have been improving substantially, but from a much lower level of development.

One criticism of rural electrification was that it might be appropriate only for regions with a relatively high level of development. That is, it would have a greater impact in rich regions, where people could afford pumps, machines and appliances, compared to poor regions, which would be unable to take advantage of electricity's many potential benefits. To make matters worse, rural electrification projects, in many cases, were implemented in the wealthiest regions first, thus increasing the possibility they would contribute to greater regional inequality.

The villages in the India study were ideal for examining the regional impact of rural electrification. The village sample was stratified according to the regional level of development of the districts, which included an advanced, an average and a poor district. Tribal regions were purposefully included in the study, making it possible to examine the various impacts of rural electrification on rich and poor communities.

Agricultural Wealth and Irrigation

The regional characteristics relevant to the impact of rural electrification on agricultural development include agricultural wealth, extent of irrigation and geographic isolation. Since more than 80 percent of India's rural population was directly employed in agriculture during the 1980s, the level of agricultural production, either per acre or per person, had a significant impact on levels of living. Value of yields in the sample areas ranged from a low of just under 100 to over 1,500 rupees per acre. These differences in yield were important for evaluating the impact of electricity on regional equity. The villages were also divided into those with high and low yields and amount of irrigated land.

Because geographical isolation generally was considered a constraint on rural development, the sample was also divided into areas with high and low levels of isolation. Without roads, grain could not conveniently be shipped outside the village to distant markets. Lack of communication may have meant farmers were isolated from new ideas and farming practices, even those being adopted on farms only 25 miles away. The advocates argued that rural electrification would help the more isolated regions catch up by more closely linking them with the larger society. However, the critics contended that the cost of connecting isolated villages would far outweigh the meager benefits derived from electricity.

Rural electrification was related to agricultural innovations in advanced regions; but surprisingly, it was related to innovations in poorer regions as well. The upward-sloping lines in figure 6.3 show that the number of years since electrification was strongly associated with innovations and high agricultural yields. *The unexpected finding was that poorer villages with low yields and little irrigation also took advantage of rural electrification.* The poor villages started out at a lower level of agricultural development, but improved just as rapidly as the wealthier ones, meaning that rural electrification was associated with innovation, not whether a village was rich or poor. Thus, rural electrification, combined with changes in farming practices, helped less developed regions improve agricultural yields.

Figure 6.3 Impact of Rural Electrification on Agricultural Innovations and Yields in India, 1981

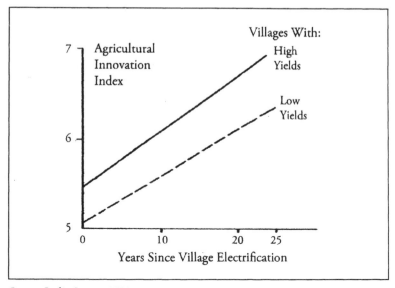

Source: India Survey 1981.

The conclusion from the analysis is that, although advanced villages did benefit most from rural electrification, *the less developed villages also improved their farming practices and yields.* Villages with low agricultural yields started out with lower levels of agricultural innovation, but improved their farming practices with each successive year of having electricity. At that time, the new hybrid seeds for non-irrigated barley sorghum and other crops were not as effective as those for rice and wheat varieties. Thus, the next section examines the patterns of development in rice- and wheat-growing regions.

Rice- and Wheat-Growing Regions

Historically, India's rice-growing regions have been more productive than other areas. With more irrigation and higher levels of rainfall, these regions have prospered relative to the country's drier areas. But during the 1970s and 1980s, the wheat-growing regions had the greatest increase in yields. Rice yields also improved over the same period, but were overshadowed by the dramatic growth in wheat yields. The lowest-yielding areas were those dependent on dryland farming for production of barley, sorghum and other grains.

In rice-growing areas during the 1980s, much of the water supply depended on unregulated streams (Sanderson and Roy 1979, p. 37), and, in many cases, waterlogging of the soil was a problem. Farmers took advantage of the remaining moisture in the soil after the heavy monsoons to produce a second crop (multiple cropping) for the same agricultural year. Thus, it was not surprising that, for rice-growing regions, the year of electrification was unrelated to the percentage of the village area under irrigation. Rice farmers relied on traditional irrigation, canals and abundant natural rainfall to produce very high yields. As a result, the year of electrification and percent area irrigated were not strongly associated in rice-growing areas.

Wheat-growing areas were at the forefront of the Green Revolution, and had dramatically improved their crop yields by the 1980s. In contrast to rice farmers, wheat farmers relied on diesel or electric pumps to improve crop production as yields of hybrid wheat varieties could only reach their greatest potential through a combination of irrigation and intensive fertilizer use. The result was that the year of electrification and irrigation were strongly correlated (0.49) in wheat-growing areas.

More surprisingly, rural electrification and irrigation were strongly correlated (0.42) in regions that grew coarse grains, pulses and sugarcane. Many of these areas traditionally relied on rainfall. Since the rains sometimes failed, the high risks discouraged investment in the new technologies. Although these regions generally had poorer yields than either rice- or wheat-growing regions, irrigation could mean the difference between extremely poor yields and moderate levels of production. In fact, being able to irrigate during light monsoon years—previously, crops would have failed during these periods—provided a buffer against the droughts that had plagued India's marginal farming areas. In addition, after irrigation, the farmers growing coarse grains typically switched to higher-value cash crops, such as cotton or sugarcane. Thus, it was an interesting twist that, during the 1980s, *rural electrification actually helped less developed regions that previously had to rely on sparse rainfall more than the already developed, high-rainfall, rice-growing regions.* By aiding less developed regions, rural electrification narrowed inter-regional disparities by providing power for much needed irrigation in India's traditionally dry-farming regions.

Rural electrification certainly has not worsened inter-regional disparities and may have lessened them. In both the advanced and less advanced villages of the 1980s, rural electrification was associated with improved farming practices. In addition, it made higher levels of irrigation possible in India's poorer dry-farming regions. With water availability, farmers switched to such higher-value crops as rice, wheat and sugarcane.

Conclusion

Both the critics and the advocates of rural electrification may have been right, in part, about the equity effects of such projects. Higher-income households generally benefited the most from the introduction of electricity in a region since they could afford the installation and monthly charges and had the ability to purchase appliances for entertainment and household work. However, the poor who adopted electricity were better off than their non-adopter counterparts. The minimum advantage for poor households was a higher quality of lighting, which improved their quality of life. In high-income countries and those with more equitable income distribution, electrification programs would also be more equitable since the benefits would be more widely spread among all income classes. As time passes, new families adopt service, and electricity use becomes quite common, even among the poorest households. Wealthy households appeared to adopt electricity almost immediately after it arrived in a village, and small farmers and lower-class households, for whatever reason, waited to obtain electricity for their homes.

Rural electrification, in combination with improved farming practices, diminished inter-regional disparities in crop production. The villages that benefited most from electricity were those with a high percentage of farmland under crops other than rice. In those villages, rural electrification was associated with irrigation and greater use of agricultural innovations, resulting in higher crop yields. In addition, irrigated farming was typically more labor-intensive than rainfed agriculture, so it was likely that the additional demand for labor benefited the low-income classes, especially agricultural laborers. The one negative finding was that having electricity caused cultivators in the poorest villages with very little irrigation potential to change their farming practices, which had not yet translated into significant increases in crop yields.

Women and children benefited more directly from having electricity in the house than men since they generally spent more time in the home. Children spent more time reading, and women had better quality lighting for working and reading. New electric appliances eliminated some of the drudgery involved in household work. Rural electrification was one of the few development programs not specifically designed to help women and children that nevertheless had very favorable consequences for them.

The short-term outcomes of rural electrification were much less favorable than the long-term consequences. In its earliest stages, rural electrification had an adverse impact on rural equality. Typically, the most populated and advanced villages were the first to obtain electricity service. Also, the wealthiest households adopted electricity in greater numbers than did poorer households, especially in the first five or even ten years of a project. For very poor countries, rural electrification programs may be inequitable initially. But over the long term, the benefits become more equitable. In Colombia, electricity adoption rates were quite high, and households of varying income levels adopted and benefited from electricity. In India, the improved crop yields resulting from rural electrification in some poor farming areas may have been enough to narrow the productivity gap with higher-yielding, rice-growing areas.

Extending electricity to the most populous and most advanced regions first was a major reason for equity problems with rural electrification. Poor, isolated rural areas generally are the last to receive electricity. However, the cost structure involved in implementing rural electrification projects is an important reason for adopting such a policy. In order to place the socioeconomic consequences of rural electrification in an economic and financial perspective, the next chapter reviews the benefits and costs of rural electrification.

7. Economics of Village Line Extension

The expense of providing rural areas electricity is an important part of the controversy over rural electrification. If electricity were extremely cheap to produce and distribute, then it would quickly replace many other types of energy, including fuelwood for cooking, diesel for agricultural pumping and kerosene for lighting. However, extending electricity to rural areas is even more expensive than meeting consumer demand in more densely populated urban regions. Power losses, long high- and low-tension lines and transformers are all quite costly. Since capital for such development projects is generally scarce, electricity distribution programs in rural areas must meet certain criteria of efficiency and cost, maximizing development impact while not creating too much financial strain on utilities.

The socioeconomic benefits of rural electrification have already been examined. The value of these benefits must be weighed against project costs to determine the price at which developing nations can afford to implement such programs. Also, the costs of grid electricity must be compared with those of alternative local systems to evaluate which technologies offer comparable service for less cost. Thus, the key components of the evaluation include the extent of the subsidy involved, whether alternative energy technologies are competitive with centralized systems, and the levels of development or conditions needed to justify rural electrification projects.

Subsidies for rural electrification are not unique to developing nations. Even for developed nations, subsidies generally were necessary in the early stages of their rural electrification programs. In the United States, for example, the relatively slow growth of rural consumer demand, combined with the high expense of rural grid extension, prompted Franklin Roosevelt in 1936 to sign into law a bill authorizing subsidized loans for rural electrification. Once electricity became a prominent part of rural life, many government subsidies were eliminated. Politically appointed commissions regulated prices to assure the electric distribution cooperatives an adequate rate of return on their investments, but also made sure prices were fair to customers.

Of course, the dramatic growth of rural demand after electrification in the United States has not been replicated in many developing nations. The dilemma for their power industry is that cost recovery depends on raising prices or improving load to make fuller use of fixed investments in plants and equipment. Raising prices may have the undesired effect of curtailing consumer demand and limiting electricity use to higher-income households. But prices that do not reflect the long-run marginal cost of producing and distributing electricity may cause shortfalls in operating capital for utilities and thereby inhibit service over the long term.

Though the benefit-cost research in this chapter was conducted in the 1980s, the lessons it offers on the complex financial issues involved in rural electrification remain valid today. The chapter reports the findings from an India case study on central-station, grid-distributed electricity, along with the cost of alternative forms of generation. Before turning to this case study, a review of the benefit-cost methods used in the 1980s for evaluating rural electrification projects is in order.

Benefit-Cost Methods

The methods used to assess the cost of rural electrification are much the same today as 25 years ago. But the techniques used to measure the benefits have changed significantly; details on the more accurate assessment methods used today are found in chapter 9. For the India case, the benefits reported in the 1980s were much lower than they would have been if calculated today. Nevertheless, they are still instructive and provide a historical baseline for methods of assessing the benefits of rural electrification.

For the 1980s case study, the costs of rural electrification were calculated in several ways and have not changed much since then. The costs always included the value of equipment for distribution schemes (construction of lines, wires and transformers). Added to these distribution costs were either (i) the average cost of generating electricity or (ii) the incremental (marginal) investment costs necessary to produce it. Most studies on rural electrification have found that it has positive economic rates of return once productive uses are taken into account. But a variety of time periods and discount rates have been used for these analyses, and assumptions concerning what should constitute costs and benefits also have varied considerably. The two main benefit-cost approaches are examined in this section.

Benefits and Costs of Rural Electrification

The financing of rural electrification projects typically is limited to distribution systems (lines, poles, transformers and meters). Generally, no specific funds are provided for generating stations and transmission. However, those costs must be included in benefit-cost calculations because, once the wires and poles are in place, households and businesses begin consuming the electricity generated from those plants. As a result, the cost of the electricity delivered to the project must be added to the capital cost of line extension for estimating overall costs.

One method for doing this used the average cost of generating electricity with existing plants and equipment. Even though this typically led to underestimating economic costs, many benefit-cost studies used this method (NCAER 1967, 1970; PEO 1965). A competing method calculated the incremental expense of supply, including the addition of new generation and transmission capacity. The justification for this latter approach was that distribution networks typically created new demand for electricity from power systems. Compared to the average cost method, this marginal cost method expressed the economic cost of power supply more accurately for developing countries with short supplies of electricity (Munasinghe and Warford 1982). The World Bank and other donor agencies adopted the marginal cost method as the standard for evaluating the financial viability of rural electrification projects, and the 1980s India case study used this approach.

During the 1980s, two distinct methods based on different assumptions about rural energy and development were used to evaluate the economic benefits of rural electrification. Both methods used customer revenue as a benefit because it was considered a proxy for minimum willingness to pay. As explained in chapter 9, this method vastly underestimated the actual benefits of rural electrification.

The first method was called the cost savings approach. It involved calculating the cost of electricity alternatives to consumers and society for providing the same level of service. For instance, the economic costs of using kerosene for lighting or diesel fuel for agricultural pumping could be compared with the long-run marginal cost (including all distribution) of using electricity for the same end use. Presumably, electricity was less expensive than the alternatives so cost savings for using it could be added to the customer revenue (willingness to pay) benefit.

The second method was called the net incremental benefit approach. Like the cost-savings approach, this method calculated total benefits by adding net incremental benefits to customer revenue (willingness to pay). Using this method, the first step was to determine that electrification was the least-cost way to provide energy for some income-producing activity. For instance, if using electricity rather than diesel for irrigation were less costly, then the total value of net incremental output due to irrigation could be considered a benefit of rural electrification. The benefit was not just the cost savings from using diesel for agricultural pumping, but the entire stream of benefits from new irrigation. All incremental economic benefits, such as improved agricultural and industrial development, could be added to household revenue to calculate a project's economic rate of return. To reiterate, the requirement was that electrification had to be the least-cost way to achieve the benefits. The 1980s India case study used the net incremental benefit approach because the country had been rapidly expanding irrigation, and many farmers, who typically kept their diesel engines until they broke down, considered irrigation with electric pumps sets less expensive.

Both approaches made numerous assumptions concerning rural energy use, especially since it was sometimes difficult to compare the same end use for different energy sources. For irrigation, one issue was whether the use of diesel pumps and engines would have even arisen in the absence of an electrification program. Another issue was the comparability of electricity and kerosene for quality of lighting. In her Indonesia survey of rural electrification, Janice Brodman found that respondents used electric lighting for longer periods of time than kerosene lamps because of the superior quality of lighting that electricity offered (Brodman 1982, p. 15). After introducing electricity to a region, power use increased because people began to use refrigerators, fans, television sets, blenders and an array of other appliances seldom powered by alternative energy sources.

Thus, providing electricity to a region may actually increase overall energy use, making assumptions concerning cost savings or even net incremental benefits somewhat difficult. Unhappiness over the 1980s benefit evaluation techniques for rural electrification led to applying more appropriate calculation techniques in the decades that followed (chapter 9). Nevertheless, the findings of the India case are still instructive for understanding the historical reasons for controversy over rural electrification in the 1980s.

Assessing Alternatives to Centralized Power Systems

Comparisons between centralized grid electricity and decentralized energy alternatives were both complicated and rather limited in scope in the 1980s. Today this issue has become even more important due to the increasing availability of decentralized electricity generation options, including micro-hydro, solar home systems and small lighting technologies. The difficulties in assessing alternatives to rural electrification are just as controversial today as they were 25 years ago.

Grid rural electrification projects are designed to last approximately 30 years, while a diesel generator has an expected life cycle of only 10 years. The panels of a household photovoltaics system are guaranteed for 20 years, but the system's batteries last only 3–5 years. Such different time frames make cost comparisons difficult (Tyner and Adams 1977; Bhatia 1979). Also, biogas systems may provide fuel for an electric generator, as well as gas for cooking. A further complication is that the amount of electricity provided by alternative technologies is often less than that of central grid service (Tendler 1979). Diesel generators may be operated only during evening hours instead of 24 hours a day. Run-of-river generation systems typically provide electric energy for part of the year, as was common in China, but in many instances such systems were unable to sustain year-round service.

Comparing these options may seem like mixing apples and oranges, but some type of comparison of the financial and economic benefits is necessary. However, the biases are fairly predictable. As a rule of thumb, comparing costs per kilowatt-hour between the generation of electricity from full-service grids and generation for part-time service may bias results in favor of central grid schemes. On the other hand, comparing costs for limited service, such as providing electricity only at night, during certain seasons or in limited amounts, favor small-scale systems. Thus, the different levels of service and reliability, along with various kinds of energy use, need to be taken into consideration during the process of selecting the most cost-efficient method for meeting village energy needs.

Rural Electrification Costs in India

The 1980s benefit-cost study in India introduced an interesting twist to estimating the financial and economic costs and benefits of rural electrification. Most benefit-cost studies use the utility or project as the unit of analysis. One unique feature of the 1980s research was that the village was the unit of analysis. The advantage was that village profiles could be developed for various combinations of benefit-cost levels, thus identifying those that were relatively expensive or inexpensive to be included in area electrification projects relative to the derived benefits. The intent was to analyze the extent of financial subsidies or profits involved in rural electrification, the types of villages for which electricity was economically justified, and how subsidies could be minimized by village selection.

During the 1980s, the costs and prices of electricity in India were roughly comparable throughout the country, so differences in benefit-cost ratios depended chiefly on the characteristics of the individual villages within projects. For the 1980s case study, benefit-cost studies were completed for 30 villages in Maharashtra, Punjab and Andhra Pradesh, thus providing a large degree of village variation for comparisons. These states were and are still characterized by distinct cropping patterns, climates and levels of development. Within each, villages differed according to their economic and social background, farming productivity, number of electric connections for agriculture, small-scale industry, commerce, household and population density, distance of village from the central grid, and types of benefits arising from the electrification project (table 7.1).

Economic benefits were more conservatively defined in the 1980s than they are today. At that time, the benefits included revenue and net incremental agricultural output, which varied greatly by village. In some villages, the benefits were limited to residential lighting, with no improvement in agriculture; while in others, cropping patterns and productivity had improved due to new irrigation pumpsets. The idea of examining rural electrification's financial costs and benefits for a wide range of villages aimed to understand how sensitive its rates of return were for various village types. This would make it possible to identify policies that would enhance the financial and economic profitability of rural electrification.

The villages producing the best financial returns in the 1980s were those with high population densities. Economic returns were greatest in villages with a high potential for irrigating one hundred or more acres of land. As was noted in chapter 6, drought-prone areas had a strong potential for increasing crops yields. These areas along with agriculturally advanced regions without extensive traditional irrigation facilities also were found to have ideal villages for including in grid rural electrification projects.

Table 7.1 Rural Electrification Characteristics of Villages in Selected States of India, 1980

Village characteristic (average)	Andhra Pradesh	Maha-rashtra	Punjab
Lines to reach village (km)			
High-tension	3.9	3.9	3.6
Low-tension	3.6	2.7	11.0
Connections in village (no.)			
Household	35.6	52.8	120.8
Agriculture	2.8	9.1	117.8
Commercial	1.7	3.5	4.0
Industrial	1.2	1.9	6.8
Net costs (US$/kWh)	0.50	0.30	0.10
Sample villages (no.)	18	8	4

Source: Venkatesan et al. 1983.

Note: Net costs are based on an exchange rate of Rs. 7.86 per US$. Although the costs appear rather high, village electricity consumption would grow over time, thus reducing the long-term net costs per kilowatt-hour.

Costs of Grid Rural Electrification

Understanding the elements involved in the delivery of electricity to villages is necessary for understanding the cost of rural electrification (figure 7.1). The generation plants that produce electricity can be powered by hydro, coal or nuclear fuel. For the 1980s study, the marginal costs of electricity from a new coal-fired thermal generating station, along with all transmission costs up to and including the regional distribution substation, were considered as the costs of generation.

Situated next to the generating plant is a step-up transmission station that increases the voltage of the electricity to about 220 kilovolts for distribution over extra-high-tension lines. The power is transmitted to a series of regional distribution substations, which lower the voltage in the lines to about 11 kilovolts and deliver the electricity through high-tension lines directly to large industry or to distribution transformers located near consumers. These transformers again lower the voltage and distribute the electricity through low-tension lines (400 volts) to rural consumers. Marginal-cost pricing assumes that all end users sharing consumption during peak demand must share the incremental capital costs involved in producing and distributing electricity from generation to customer delivery.

Figure 7.1 Diagram of Typical Grid Distribution System

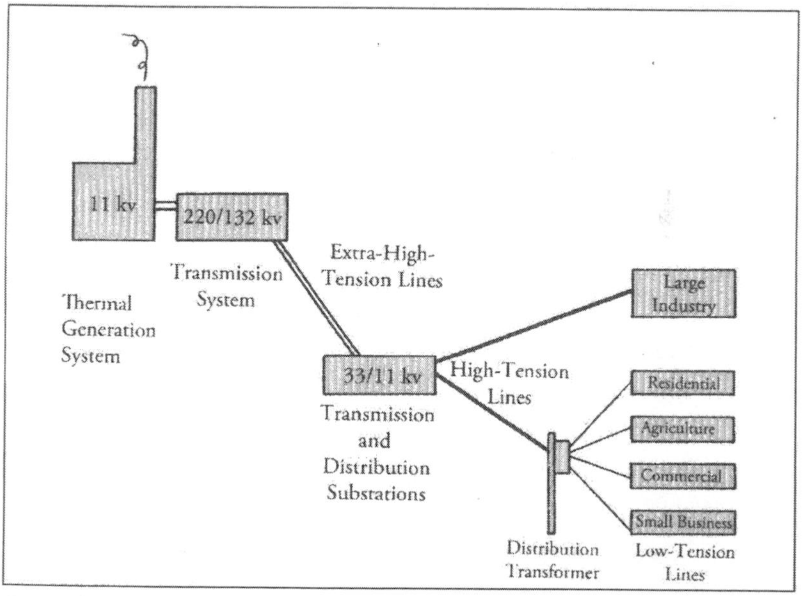

Source: Barnes 1988.

A two-step process was used to estimate the total cost of delivering electricity to villages for low-voltage consumers. The first step calculated the marginal cost of bringing electricity to the distribution substation near the village. This cost included the capital and operating costs of generation, transmission and subtransmission. The same cost for delivering energy to the distribution substations, based on total kilowatts generated in the system, was assumed for all villages in the study. The second step calculated the marginal cost of distributing electricity from the distribution substation to each village. This cost varied according to village characteristics, including population density, distance from the substations and customer type.

In the 1980s, the marginal cost of electricity for rural electrification consisted mainly of the costs to extend lines to the village and provide local service (Venkatesan et al. 1983, pp. 26–32; Barnes and Jechoutek 1984, annex A). The cost of distributing electricity to each village depended on the length of high-tension (11 kilovolt) and low-tension lines, as well as the number and type of connections (Venkatesan et al. 1983, pp. 70–71). The implication was that villages located farther from the grid system and with scattered households were more expensive to serve because of the longer lines ultimately needed to reach rural households. Thus, to obtain the total cost of providing electricity to each village, the cost of bringing electricity from the distribution substation to the consumers of a particular village had to be added to the average marginal costs of generation and transmission.

The cost to reach a typical substation in the vicinity of a was about 14 percent of the total costs of providing electricity (table 7.2). The 1980s long-run marginal cost for generating and delivering energy over high-tension lines was about US$0.02 per kilowatt-hour. In a typical village, the cost of distribution was almost six times the cost of generation and high-tension distribution. The reason is that rural consumers were generally geographically dispersed, requiring extensive investments in lines, equipment, and maintenance per customer. For all the villages in the study the average cost of medium tension lines to the village was US$ 0.04 per kWh and low-tension village lines were the same. Transformers accounted for $0.01 per kWh. The remainder of costs involve energy losses, operation and maintenance for a grand total of $0.13 per kWh.

Table 7.2 Long-Run Marginal Cost for Typical Village in India, 1980s

Costs involved in extending grid to typical village	1981 percent of costs	1981 Costs (Cents)
Generation and high voltage transmission	14	1.8
Generation, including energy cost	12	1.6
Transmission and subtransmission	2	0.2
Village costs	86	11.2
Operation and maintenance	14	1.8
Low-tension lines	31	4.0
Transformers	7	1.0
Medium-tension lines to village	31	4.0
Energy loss in distribution	3	0.4
Total	100	13.0

Source: Venkatesan et al. 1983.

Note: Total 1980 net present value cost was about Rs. 2 or US$0.13 per kWh.

Of course, for specific villages the costs are different from the average of all villages. For instance, one village in the study with 1 commercial, 30 household, 5 agricultural, and 2 industrial connections had a yearly electricity use of about 18,490 kilowatt-hours. For this village, the 1980s cost per discounted kilowatt-hour for 2.75 kilometers of high-tension lines, 2.75 kilometers of low-tension lines and 1 transformer, along with facilities operation and maintenance, totaled US$0.20. After adding the generation and high transmission marginal costs, the grand total for the long-run marginal cost was approximately $0.22 per kilowatt-hour. This was the estimated marginal cost of electricity generation and transmission and the full marginal cost of distribution from the substations to the village.

The relationship between consumer density and electricity load profile was extremely important for the per kilowatt-hour cost of grid electricity (figure 7.2). For villages in the study, the cost per kilowatt-hour ranged from US$0.12 in Punjab, which had extensive connected load, to more than $0.62 in villages with low load and extensive lines.

These results from the 1980s study assumed increasing electricity use for the first five years, based on actual connections in the village, but no growth for subsequent years. For individual villages, the cost per kilowatt-hour depended heavily not only on the length of high- and low-tension lines to the village, but also on the number of consumers adopting electricity. Villages with high electricity adoption rates were much less costly—an argument for promoting universal electricity adoption in villages with electricity. Even after 25 years of having had electricity, many of India's villages still do not have universal adoption, which raises the costs of village electrification.

Figure 7.2 Cost of Rural Electrification by Load and Customer Density

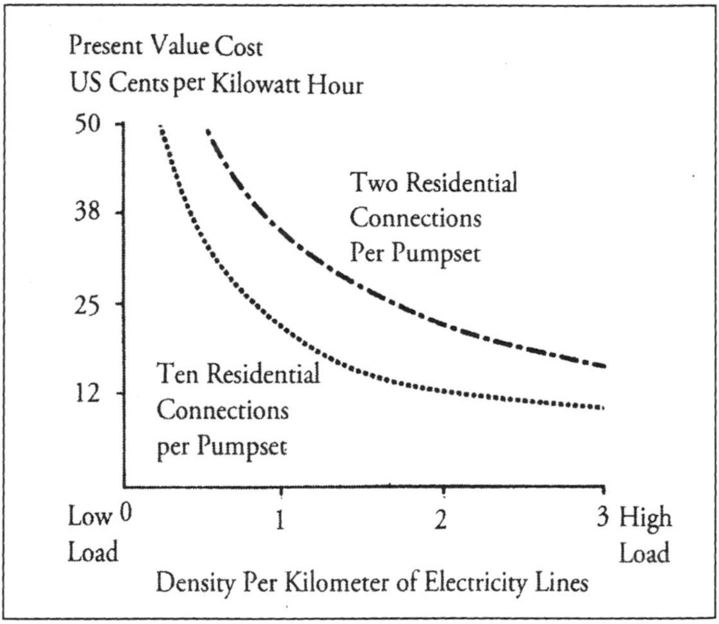

Present Value Cost
US Cents per Kilowatt Hour

Two Residential Connections Per Pumpset

Ten Residential Connections per Pumpset

Low Load — High Load

Density Per Kilometer of Electricity Lines

Source: Barnes 1988.

The general rule of thumb in the 1980s was that at least one or two pumpsets and a significant number of household connections per kilometer of lines were necessary to keep the cost of electricity below US$0.25 per kilowatt-hour. The study's conservative assumption was that villages would reach maximum levels of connections after only five years. The conclusion is that the conservative growth estimates somewhat overstated kilowatt-hour costs, especially for villages with low loads, few connections, and high cost per kilowatt-hour. People in these villages eventually would see the value of having electricity, and no doubt would have added appliances and small machines to make life easier and more productive. The consequence would be lower electricity costs than assumed for the 1980s study.

Comparative Cost of Irrigation Pumping

At the time of the 1980s study, the comparative cost of diesel versus electric irrigation was mixed. For instance, a report by the Swiss Development Corporation (summarized in Bhatia 1983) concluded that diesel pumping would be less costly than electricity. On the other hand, the Administrative Staff College of India (1980, p. 121) concluded that electricity would be less costly than the diesel alternative.

During the 1980s, many project appraisal documents used the net cost approach as a benefit for electricity replacing diesel pumps. Thus, it was important to assess which technology was least cost for irrigation. Diesel and electric pumps provided approximately the same benefit in supplying water for irrigation and thus improving crop yields. The main exception was for deep tubewells and dugwells, where diesel engines were not practical because access to the pumps would be difficult or impossible. However, it was possible to compare the cost of electric diesel pumps for dugwells to determine which alternative would be more economical for agricultural pumping.

Assuming that electrification projects were multipurpose and electricity was extended to all consumer classes, the long-run marginal cost for villages was in the range of US$0.12–0.60 per kilowatt-hour. By comparison, the capital and operating cost of diesel engines was about $0.26 per kilowatt-hour. Thus, the conclusion was straightforward: Electric pumps were less expensive in densely populated areas, especially those closer to the distribution stations, but more expensive in more remote villages with few farms and little irrigation potential. Just under half of the villages surveyed in the 1980s had lower kilowatt-hour costs than diesel. This was especially true in the Punjab, where agricultural pumping was extensive and the average marginal cost of electricity was $0.10 per kilowatt-hour. However, at the time, the subsidized price of electricity for agricultural pumping was less than $0.02 per kilowatt-hour, so it is little wonder that farmers perceived electricity as the least expensive way to irrigate their fields.

Community Power Generation and Distribution

The distribution of electricity powered by medium or small generation systems was an option in rural communities even 25 years ago. Such power sources as micro-hydro, mini-hydro and diesel generators were alternatives to centralized grid electricity. Mini-hydro had become widespread in China. Even the United States saw a substantial increase in decentralized generation systems following the enactment of legislation requiring utilities to purchase power from privately owned, decentralized alternatives (Flavin 1986).

Local conditions are extremely important in determining the comparative costs of central rural electrification and decentralized systems, such as diesel or other forms of local generation. Grid rural electrification projects are favored when 24-hour service is needed and electricity is used for multiple purposes, including agriculture, commerce, industry and residential consumption. The central grid is also competitive when there are many connections per mile of electric lines and the load for the overall system is well-managed.

The conditions that favor smaller, decentralized electricity service include part-time service (e.g., evenings only), single end use (e.g., irrigation pumping) and low energy demand. If agriculture is the only end use, electrification is not viable in situations where farm plots are distant from the central grid. However, the incremental cost of providing electricity for agriculture would decline if electricity were extended to villages for many purposes. The other services would share the fixed costs of extending electricity to a village. Given the complexity of these assumptions and local conditions, it is understandable that different conclusions were reached concerning the superiority of central grid electricity or decentralized forms of electricity for community systems.

One persistent problem in developing countries even today is that decentralized systems often are not compatible with the central grid because of voltage levels and non-standard equipment. Generally, decentralized small- and medium-scale generation is most cost effective in situations where electricity can be generated for both the community and sale to the central grid. Decentralized generating units that provide additional energy to the system during peak-load periods (e.g., micro-hydro) would be especially attractive. In remote regions, small- and medium-sized decentralized systems can be cost competitive with central station generation and should be considered as alternatives when local resources are available. Thus, even in isolated rural areas, standardization of systems for decentralized generation is important to avoid having to dismantle and replace them once grid electricity arrives in the region.

Micro and Pico Household Systems

The options for single household electricity systems were quite limited in the 1980s. In many rural areas today, connecting to the electric grid still may be decades away and even then may be financially prohibitive. But new alternatives to grid electricity and carbon-based fuels mean that households do not have to wait for the expansion of grid electricity systems (table 7.3).

A wide array of new and renewable energy systems can provide for both specific end uses and general rural energy services. There is thus a possibility of accelerating the transition to modern energy services through the expanded use of off-grid renewable electricity systems.

Moving from grid-based electricity to off-grid options can increase the kilowatt-hour cost. For instance, grid systems can provide electricity for as little as US$0.10 per kilowatt-hour, compared to $40 per kilowatt-hour for a battery-powered flashlight, which is quite costly. Generally, electricity becomes more expensive for people in remote regions, particularly for those who are among the world's poorest populations. Nevertheless, smaller electricity systems involving lower upfront costs can be sensible options for people in remote areas.

Table 7.3 Electricity Provision Options for Rural Households

Electricity source	Household capacity	Possible benefits of electricity service
Main grid electricity	1 kilowatt and above	Lighting, color TV, black-and-white TV, cell phones, small appliances, fans, refrigerators, air conditioners, motors and electric pumps
Isolated grid electricity	500 kilowatt and above	Lighting, color TV, black-and-white TV, cell phones, small appliances, fans, refrigerators, air conditioners, small motors and electric pumps
Photovoltaic household system	15+ watt-peak, coupled with 700 watt-hours battery storage (60 Ah)	Lighting, color TV, black-and-white TV, cell phones and low-wattage small appliances
Car battery	700 watt hours of battery storage capacity (60 Ah)	Lighting, color TV, black-and-white TV, cell phones and low-wattage small appliances
Small/pico lighting system	5–15 watt-peak, coupled with various small batteries	Lighting, cell phone charging and radios
Battery-powered flashlight	D Cell batteries with 2–6 watt hours of storage capacity	Lighting, cell phone charging and radios

Source: Barnes 2010.

Benefits of Rural Electrification

In the 1980s, traditional benefit-cost analysis often concluded the potential benefits of rural electrification were not worth the significant infrastructure costs required. It was acknowledged that rural electrification improved quality of life and economic produc-

tivity, but expansion to rural areas was costly because of remote locations and the dispersed nature of beneficiary populations. While
such benefits as better health, improved education, increased
productivity and more leisure time were often mentioned, they were
then disregarded as too hard to measure. With little relevant information to judge the impact of those intangible benefits for rural
populations, political decision makers were left in somewhat of a
quandary about whether to fund projects. In this section, some of the
benefit estimation techniques from the original 1988 study are sup-
plemented by more recent research.

Revenue-Based Benefits

During the 1980s, evaluations of rural electrification's benefits
and costs considered power company revenues as both financial and
economic benefits. Because revenues were considered the minimum
willingness to pay, financial and economic returns were almost the
same. For most of the villages surveyed in India, the price charged
for electricity per kilowatt-hour was approximately US$0.02 for agri-
cultural pumping and about $0.06 for residential, commercial and
industrial uses. At the time of the 1988 study, the marginal cost for
an average village was about $0.26 per kilowatt-hour, many times
higher than the revenues collected by the utilities (table 7.1).

For the electricity companies, rural electrification was a losing financial proposition unless they could recover some of the costs
through government or state subsidies or higher electricity prices.
For villages covered in the 1988 study, losses to the distribution
companies averaged about US$0.24 per kilowatt-hour for agricultur-
al pumping and $0.20 per kilowatt-hour for residential use. As a re-
sult, the utility companies were hesitant to further expand or finance
rural systems because each new customer meant they would lose
more money. These companies received both state and national subsidies for extending distribution facilities to villages, but little attention was paid to promoting full coverage in villages that were already
connected. The drastic difference between the costs and revenues
may well have been the reason for the poor reliability and service
experienced in rural areas of India even today.

Before making any judgments on the efficacy of rural electrification program, it must be determined whether their economic benefits justify the economic costs and under what conditions they could be profitable or at least incur minimal losses. The revenue-based benefits are more typically understood as financial flows for an electricity company's balance sheet rather than economic benefits for the country. After years of refinement, revenues are still considered financial benefits for rural electrification projects, but the techniques used to measure economic benefits have changed significantly since the 1980s due to better information about rural electrification's impact on development (chapter 9).

Production Benefits

In the 1980s studies, the production benefits of rural electrification were considered the result of changes in irrigated land. The farmers surveyed believed that the use of pumpsets was essential, not only for irrigation during the main cropping season, but also as a precaution against possible failure of the monsoons. During critical periods of crop growth, significant drops in precipitation could lead to near total financial ruin for farmers. For many regions of India, the net production benefits of irrigation were substantial. At that time, some farmers benefited by switching from growing groundnuts to intensive rice cultivation. Other farmers produced more rice or wheat using improved double- and triple-cropping techniques. In cotton-growing regions, the improved yields were substantial even when only a small number of acres was affected.

Comparisons between villages in Punjab and Maharashtra in the 1980s indicated that, not only was it important for rural electrification to have an impact on an irrigated area; it also needed to have an impact on agricultural productivity. Low productivity per acre in Maharashtra meant that, even for villages with 100 irrigated acres, electrification provided significantly fewer benefits than in the higher productivity villages of Punjab. For instance, one village in Punjab had a negative financial return of about US$127 until adding agricultural benefits, after which it had a positive economic rate of return of more than $250,000. In Andhra Pradesh, similar villages with canal irrigation that were provided electricity had virtually no pumps or incremental agricultural benefits. This is similar to the experiences of Indonesia, where canal irrigation inhibited growth in the use of electric pumps for irrigation.

Village Profitability and Subsidies

According to the 1988 study, the cost of extending electricity to India's villages was greater than the revenues received for most villages. For all villages in the study, the financial revenue–based, net present values were negative, an artifact of artificially low electricity prices. The variation in the subsidy to the consumers was high, ranging from a negative net present value of US$700 for villages with the lowest subsidies to $125,000 for the highest subsidies. However, production benefits were, in many instances, quite substantial, which meant that the projects perhaps could be economically justified. The wide fluctuation in returns indicates that some villages benefited more than others; these findings are in line with the earlier rule of thumb developed for the cost of rural electrification.

As might be expected, the financial viability of rural electrification improved for villages with a high number of customers per mile and a small number of kilometers of low-tension lines. For five villages located near the distribution substations, the project's financial subsidies were relatively low. Even two villages with scattered consumers requiring extensive low-tension lines but short high-tension lines from the grid required lower subsidies than expected. Less obvious, villages in areas with an intensification program through which nearly all households adopted electricity incurred minimum losses. Such villages also shared transformers with nearby settlements. Thus, the financial viability of projects was greatly affected by the physical layout of the distribution system and the village connection load (connections per kilometer of lines). To minimize losses, design of the distribution systems should be based on well-understood patterns of expected rural demand for electricity (Sen Gupta 1984; Munasinghe and Warford 1982).

Another obvious factor affecting a system's financial viability was the price of electricity. As previously indicated, the price for agriculture was one-third of what households paid. Consequently, villages with high numbers of agricultural connections generated more financial losses for the distribution companies, even though the economic returns for society were attractive. For instance, villages in the Punjab with more than 100 acres of land irrigated by electric pumps generated high financial losses, which were more than offset by the high rates of economic return from the production benefits. In addition, the practice during the 1980s of appraising the economic benefits of rural electrification through revenues as a proxy for willingness to pay meant that low prices caused a significant underestimation of the benefits.

Social Benefits

The social benefits of rural electrification in the 1980s study were never that well-understood and were not quantified in economic terms. The benefits included higher quality of rural life, more reading time by children, higher impact on women and children than men and the use of new appliances in the home. Other benefits that were often ascribed to rural electrification were not proven. These included diminished rural-to-urban migration and reduction in birth rates. Since that time, availability of jobs and education, rather than rural electrification, have been found to be more closely related to rural-to-urban migration, and there has been mixed evidence on birth rates. The emergence of growing towns and service centers that retain rural populations may be affected by rural electrification, but even now, the evidence is not complete. The only negative benefits of rural electrification included an adverse effect on rural equity, especially early in the programs.

Such positive benefits as improved education and use of labor-saving electric appliances are difficult to quantify in monetary terms, but should be included in the social benefits of rural electrification. Recently, techniques adopted from both traditional and environmental economics have been used to develop ways to quantify in monetary terms some of these identified, hard-to-measure benefits. The results of these new studies confirm that the economic benefits of rural electrification are quite large and much higher than the costs involved in extending lines to rural communities (chapter 9). For instance, the amount of light from an electric lamp can be 50–100 times higher than that provided by kerosene lamps; because the higher-quality electric light also is much cheaper, households use more of it. Television viewing also is much cheaper when powered by grid electricity rather than a car or motorcycle battery. The value of these benefits can be quite high, ranging from US$15 up to $40 per month. However, recovering adequate revenue to maintain lines and quality of service remains a problem in India to this day.

Conclusion

Rural electrification programs should be considered part of an overall commercial energy strategy. The general trend is that, as societies develop, traditional fuels decline as a percentage of total energy use since new uses are found for commercial energy. For regions with the necessary resource endowments, mini-hydro can be a low-cost alternative to complement the grid electricity generated from central stations. Conservation of existing energy without diminishing rural productivity also is important for rural electrification since this avoids expensive investments in plant and distribution facilities and alleviates the power shortages experienced in many developing nations. Many types of energy systems can be advantageous in providing commercial energy to rural areas distant from the electricity distribution substations and villages with low levels of electricity demand.

In densely populated regions close to the national grids, rural electrification is generally less expensive than photovoltaics or wind power with battery storage. For those villages and pumpsets located farther from the distribution networks, diesel engines and biogas generation can be cost competitive with central grid-based rural electrification. Small wind-energy systems may only be appropriate in those regions with the necessary wind resources. Nevertheless, villages with low electricity needs may well benefit from the development of such alternatives as biogas, photovoltaics, wind and community generation.

The main factors affecting the financial and economic feasibility of rural electrification schemes include (i) load development and management, (ii) layout of the distribution system, (iii) extent of economic benefits resulting from the projects and (iv) electricity pricing. Load development and management are important in several respects. Greater electricity consumption per mile of line means a lower unit cost of providing service to rural areas because of the better use of fixed investments. Placing distribution transformers and lines in rural areas and using them for agricultural pumping only two or three months out of the year would mean the hardware is underutilized most of the year. The development of multiple uses for electricity in rural areas is highly desirable from the standpoint of central grid electrification, so complementary rural development or productive-use programs are quite important. Residential electricity use occurs year-round, but mostly during evening hours. Agricultural processing and water-lifting loads occur at different periods of the year, but the load itself occurs mostly during the daytime. Many commercial establishments opened year-round draw small amounts of electricity during the day and evening hours. By coordinating the use of electricity for different end uses, the load factors are improved and the cost of electricity per kilowatt-hour declines substantially.

The layout of the distribution system is quite important to the overall cost of electricity in rural areas. Not only are the distribution lines expensive; longer lines also mean greater technical losses within the system. In some countries, planning of the distribution network is viewed as putting the hardware in place. But to minimize system costs, the potential connected loads for agriculture, industry and residential use should be assessed through socioeconomic surveys before planning the distribution system layout. Perhaps even a minimum number of connections or load levels should be firmly established prior to extending lines and placing transformers in rural communities. To be cost competitive with diesel, about one electric pumpset is necessary for every kilometer of distribution lines. In regions where such pumpset densities can be achieved, the economic benefits of irrigation for farmers and society are substantial. Despite the high capital costs involved in distribution systems, advance planning can lower line costs and electricity losses considerably, with virtually no impact on quality of rural service.

In the 1980s, the social benefits of rural electrification, though highly valued by rural consumers, had not been adequately factored into the benefit-cost evaluations, primarily because they were hard to quantify in monetary terms. At that time, the social benefits were weighed along with other factors in determining rural consumers' willingness to pay for electricity. On the other hand, prices were set for a variety of reasons, and, in the case of India in the 1980s, reflected minimally on consumers' willingness to pay. Most project assessments in the 1980s vastly undervalued rural electrification's social benefits, an issue dealt with further in chapter 9.

In summary, the least-cost energy alternative for the same level of rural service is probably the best method for evaluating the viability of rural electrification. Such an approach broadens the evaluation to define the most appropriate rural energy strategy for specific communities or regions. In general, an improved knowledge of present and predicted future energy use is necessary before rural electrification projects become the most economical way to meet specific rural energy needs. Under the right conditions, grid forms of rural electrification can be least-cost energy alternatives for socioeconomic development, but sparsely populated regions with little potential for productive use are the wrong sites. Such remote areas are better serviced by commercial decentralized energy systems. The right regions for grid rural electrification include densely populated areas—especially those with potential for productive uses of electricity—both rich and poor agricultural regions with underutilized groundwater resources and areas with relatively high levels of income.

8. The Continuing Debate

Rural electrification is an important component of energy strategies for developing nations. The 1980s criticism of rural electrification helped to focus attention on both the previous assumption that electrification was a positive force in rural development and the policy issues involving the most appropriate energy strategies for socioeconomic growth. As is usually the case in most controversies, the outcome was neither black nor white; in some cases the criticism was justified, while in others it was not. Portraying rural electrification as a universal solution to rural energy and development problems certainly was incorrect. But the position that electricity projects contributed virtually nothing to development was equally fallacious.

This research conducted during the 1980s found that rural electrification had more of an impact than had been anticipated at the outset. Nevertheless, central grid distribution was not always the least-cost energy alternative for rural development. Obviously, electricity was not the only energy program that should have been considered for all situations in developing nations, yet it certainly was an important component for rural energy development. This period marked the very beginning of the acceptance of alternatives to central grid electrification, which in time would come to include household photovoltaic systems, mini- and micro-hydro community grids and small lighting systems.

The more recent emphasis on sustainable energy for all is, in part, an outcome of the movement toward development of decentralized solutions to the rural energy problem. The danger is that, if the public policy pendulum swings too far in the direction of providing the most expensive energy to the poorest rural populations, the long-term achievement of energy for all might suffer from the same types of criticism as the 1980s programs. Thus, the lessons learned during the 1980s concerning the role of rural electrification systems for development are just as relevant today as they were 25 years ago.

Electricity for Rural Development

The evaluation and resolution of the policy issues in the 1980s rural electrification controversy involved assessing improvements in rural productivity, determining the social benefits and weighing the economic benefits and costs. In the beginning of the research, several important questions were asked concerning the efficacy of rural electrification projects. Many of these questions were answered in the process of the research, which also raised new questions.

Agricultural Productivity

In the 1970s and 1980s, India's policy of stressing the importance of electricity for rural development led to a greater agricultural impact than was experienced in Colombia or Indonesia at the time. Rural electrification definitely enhanced India's agricultural development. In addition, India implemented an aggressive agricultural development program known as the Green Revolution, which included the dissemination of hybrid seeds, fertilizers and other agricultural inputs. Complementary credit programs also were beneficial in that they allowed farmers to change from traditional to more modern farming practices. Indonesia had no similar credit programs that would aid farmers in obtaining agricultural pumpsets or developing small-scale businesses.

Rural electrification has had an impact on India's agricultural development through use of irrigation pumps and new agricultural technologies. In India, the price of electricity for agricultural pumping was subsidized heavily to both good and bad effect. The subsidies demonstrated the government's commitment to rural development, but also made it difficult for the electricity companies to recover their costs—problems that persist even today.

During the 1980s, the impact of rural electrification occurred through the following causal chain. Electricity lines were extended from the central grid to villages. With access to electricity, farmers replaced diesel engines or purchased new electric-powered pumpsets. This process did not occur immediately; rather, it was slow and continuous. As more pumpsets were added in a village, a higher percentage of village farmland was irrigated, which led to higher cropping intensity and farming land for more than one season. It also allowed farmers to use agricultural inputs, including hybrid seeds, fertilizers and pesticides. Growing a second crop increased farmers' income and required more agricultural labor for land preparation and utilization. As a result, both farmers and agricultural laborers reaped the benefits of irrigation. The improvements in irrigation, cropping intensity and use of agricultural innovations led directly to higher agricultural productivity.

This model of how rural electrification related to agricultural productivity should be qualified by the knowledge that some regions benefited more than others, and diesel pumps had about the same impact as electric pumpsets. The areas that benefited the most from rural electrification were not the relatively rich, traditionally irrigated rice regions with heavy rainfall; rather, they were the drier and, in many cases, poorer areas with sufficient but relatively untapped groundwater sources for irrigation. The wealthier wheat-growing regions also benefited substantially from rural electrification. Interestingly, the traditional forms of irrigation, such as wells with Persian wheels and tanks, were not associated with improvements in agricultural innovations or higher crop yields. Also, the growth of pumps was slowest in those regions with extensive canal irrigation systems. One conclusion, therefore, was that rural electrification helped to improve regional levels of equity through the development of farm-level irrigation.

India's effort to improve rural development through electrification was successful, but no parallel impacts were experienced in Indonesia or Colombia. One reason for the lack of impact in the Indonesia case was that the survey was conducted in a traditionally irrigated, rice-growing region. Even in India, the growth rate of electric pumps for irrigation was low in such regions. Also, Indonesia heavily subsidized the price of diesel fuel, which diminished the attractiveness of electric pumps for farmers. In Colombia, the farming systems were not as intensive as those in the Asian countries, and the growing seasons were somewhat shorter. As a result, Colombia had less demand for water lifting, but there was still scope for irrigation beyond existing levels. Some parts of Colombia probably could have benefited from a more extensive agricultural development program that would have featured irrigation, fertilizers and hybrid seeds. That country also is reliant on export plantation crops, and the coffee-growing region utilized electricity for lighting and coffee-processing activities.

Today many of the world's remaining areas without electricity could benefit from the lessons of India. To be sure, the situation in Africa today differs substantially from the case of India 25 years ago, but the emphasis on supporting small-scale farmers' access to electricity, credit and complementary rural development inputs could make a big difference in moving Sub-Saharan Africa into the 21st century. It should be remembered that India during the early 1970s experienced dismal yields and food shortages. The response of providing the necessary inputs for agricultural development is a lesson that should not be forgotten.

Small-Scale Industry and Commerce

Businesses in rural areas of India, Colombia and Indonesia typically included small commercial shops, grain mills, sawmills, and brick-making. In all three countries, rural electrification resulted in additional productivity either through improvements in existing industries or by encouraging the development of new businesses. Rural industries using electricity in India, most typically grain mills, had greater labor productivity, lower fuel cost as a percent of total costs, employed more workers and were larger than industries without electricity. Virtually all industries in India adopted electricity when it became available. In Colombia, the benefits of electricity for rural businesses were lighting, refrigeration and a few driveshaft power applications. In Indonesia, because of the low price of diesel fuel and high connection charges, businesses seldom used electricity for driveshaft power. However, because of electric lighting and refrigeration, businesses stayed open longer, employees worked more hours and productivity improved. The conclusion from these legacy studies was that electricity was beneficial to businesses in all three countries.

Electricity was just one of the important inputs that helped in the development of rural industries. The other complementary conditions included access to good rural markets, adequate credit and the entrepreneurial skills of business leaders. Extending electricity to regions without paying attention to these other conditions would lead to lower-than-average growth of rural businesses. Without sufficient markets for their goods and services, businesses would have a difficult time. Credit also may have been necessary for start-up capital or expansion of existing industries. Perhaps because these complementary conditions were not present in all rural areas, the anticipated growth of rural industries with electricity was somewhat slow. However, the areas without electricity had an even worse record of business development. Thus, the conclusion from the 1980s studies of rural businesses was that electrification was an important condition for their development, but an explosion of industry and commerce should not be expected.

Even today, the extension of electricity to areas previously without it generally is the responsibility of energy companies. The attention to complementary conditions for the development of rural enterprises and businesses is not part of energy company mandates. The possibilities of using electricity for rural business development are quite varied; yet, in the rush to provide electricity to rural people, the productive uses of electricity are sometimes forgotten. The main reason is that, unlike the "electric circuses" that promoted both appliances and electricity equipment for productive uses during the rollout of the U.S. rural electrification program, a main emphasis of today is energy conservation. Fortunately or unfortunately, for the development of rural areas, promotion of electricity use should take precedence over conservation, although the use of more efficient equipment also should be encouraged.

Social Impact and Equity

The social impact of rural electrification was greater than anticipated. At the time, many studies had reported that electrification resulted only in a few light bulbs being used in the wealthiest households. However, the 1980s research found this presumption widely off the mark. Electricity was used for both lighting and appliances. In fact, household appliance use was greater than expected, especially in Colombia, where a high percentage of households with electricity had acquired television sets; some were even cooking with electricity, and others had refrigerators. Appliance ownership resulted in a change in evening social patterns, whereby persons tended to spend more time watching television and less time visiting or engaging in productive activities. However, children in households with electricity were able to study more, and this was true for both India and Colombia. A higher than expected number of households in rural India purchased fans, which are important for quality of life in a hot climate. Later studies in India's rural areas confirmed that the use of appliances would grow over time, and today televisions are quite common. Supporting this notion, households with electricity in India reported having a generally higher quality of life than those living in households without electricity, and this was especially true for middle-income groups.

Migration and education were affected by rural electrification in quite different ways. Electric lights, which are superior in quality to kerosene lamps, make it easier for children to study during the evening. On average, even among households at the same income level, those with electricity also had higher levels of education, so the attractiveness of electric lights no doubt was responsible for the strong association between literacy, education and rural electrification. More recent findings have confirmed that rural electrification is an important cause of higher school-enrollment rates, no doubt because children can study more during the evening hours.

Contrary to the expectations of most analysts, migration increased because of rural electrification. Many advocates of rural electrification had expected that improving quality of life in the countryside would reduce rural-to-urban migration. In fact, villages with electricity in India appeared to experience greater migration to urban centers than villages without electricity. However, this finding was consistent with expectations that more, rather than less, migration should accompany rural development. Young people seeking opportunities beyond the farm moved for education and jobs. As job opportunities increased in rural communities, a significant amount of in-migration took place. The Colombia results indicated significant migration to the coffee region—areas with high levels of commercial agriculture and electrification—for employment. Also, people tended to stay longer in regions with electricity, a finding that supports the notion that electricity stems out-migration. Thus, the results for migration appear to have depended on local circumstances surrounding regional development. Little new research on migration has been conducted since these 1980s studies. Until more recent study results become available, these legacy findings remain among the few studies on the relationship between rural electrification and migration even today.

It was found that rural electrification had a negative effect on inequality over the short run, but improved it over the longer term. When electricity first arrived in rural communities, wealthier households had the money and influence to obtain a connection before anyone else. Thus, wealthy households had access to the considerable benefits made possible by electricity, while poor households were left in the dark. Television, fans, radios and lights all improved the quality of rural life for those who could afford electricity. As time passed, however, more and more households were able to adopt electricity, and the programs became much more equitable. One interesting positive impact was that rural electrification typically benefited women and children more than men, since men spent less time in the home. In fact, more recent studies have confirmed that rural electrification has significant household benefits, especially for women and children.

Rural electrification did not cause greater land concentration or rural poverty in India. Large farmers were not buying out smaller farmers, who could not afford to invest in pumps and new technologies. Electrification appeared to be neutral with regard to farm size, although there may have been some consolidation of land parcels to take advantage of electric or diesel pumps. This neutral impact may have been the result of the land ceiling laws, which discouraged India's large farmers from acquiring new land. In addition, rural electrification did not appear to have been a cause of rural poverty. If anything, it may have helped lessen the percentage of villagers living below the poverty line, probably through the additional employment of agricultural labor. More recent studies confirm that rural electrification seems to lessen the number of people in poverty; that is, over the long term, the socioeconomic benefits of rural electrification appear to outweigh the short-term negative equity impact.

Thus, the more recent studies on the social impact of rural electrification confirm the results of these 1980s studies in India, Colombia and Indonesia. The more recent research is important for clarifying the role of rural electrification in development and adding to the evidence that providing energy for all is an important contributor to development outcomes. But the fact that rural electrification, in conjunction with complementary conditions, not only has greater impacts but also leads to higher revenues and better financial conditions for the electricity companies is a message that still has not received enough attention.

Costs and Benefits

Nearly all rural electrification schemes involve financial subsidies during the initial stages of rolling out systems to rural areas. The size of the subsidies depends on the population density of the village, the agricultural acreage affected by electric pumps, the expected adoption rate of electricity and such regional factors as climate and level of development. In many cases, the economic benefits of additional agricultural and industrial production balance the extensive financial subsidies necessary for rural electrification projects. That is, the overall benefit to society is greater than the cost. However, even if the socioeconomic benefits are high, cost recovery by the power industries may be quite low or negative, causing them to depend on large government loans or grants to maintain service. The restricted financial viability of rural electrification is one reason why power industries previously considered rural service a low priority.

The wide variety of rural energy options available today does not change the economics of rural electrification. In fact, financing rural electrification is even more difficult today than it was 25 years ago. The reason is that the world's remaining 1.3 billion people without electricity are often located in poor, remote regions. The costs of serving these regions with electricity are even higher now than previously. Solar home systems and decentralized grids are quite expensive. Thus, the old adage, "renewable energy is free, but it is not cheap," still applies. The cost of serving those who remain without electricity will be a challenge for both grid rural electrification and decentralized alternatives for producing electricity.

Important Policy Considerations

The role of rural electrification in development can be quite important if complemented by the right supporting programs. Because electricity can be used for multiple purposes, ranging from lighting, communications and space conditioning to water pumping and productive work, government policy plays an important role in shaping the impact of rural electrification. A greater diversity of electric appliances and machine use results in greater impacts for rural areas, so narrowly focused projects should be avoided. The ambitious attempt to spread electricity across an entire nation too fast and too soon may lead to severe financial strains on the power industry, and may hinder such programs in the long run. Electricity programs are costly; although huge financial losses are possible, they can be avoided by having adequate pricing policies, complemented by careful planning for load development. The extension of electricity to villages should be coordinated with other development programs, as electricity requires complementary inputs to have significant impacts on development. In general, energy planning needs to be forward-looking, anticipating future growth and development.

Rural electrification is a capital-intensive investment. If developing countries are not careful, its tremendous costs and operating subsidies can divert capital from other much needed development programs. Long-term reliable electric service is an important goal that can be assured if financial subsidies are minimized by concentrating on villages with significant loads and good planning to minimize the length of high- and low-tension lines. In villages with adequate load per kilometer of lines, electricity will be less costly per unit of energy and have a greater impact on the community. Assuring the financial health of the power industry will mean that system expansion will not create severe strains on the reliability of rural electrification service.

Not surprisingly, the villages that have demonstrated the most significant growth in electricity use are those with large populations. For residential electricity, conditions conducive to load growth include larger populations, high literacy rates, greater access to roads, adequate credit availability and high agricultural income. Although this would suggest that perhaps only the most developed villages should receive priority, many poor villages have also benefited substantially from rural electrification, especially those with adequate groundwater for irrigation. The 1980s studies found that in rice-growing regions, where there is little demand for agricultural pumpsets, residential service should probably receive the highest priority. For promoting the development of rural businesses, the availability of markets, credit and rural literacy were associated with growth of electricity use. The fact that literacy spurs household, business development and agricultural innovations shows that education and rural electrification are quite complementary programs.

So far these policy guidelines for rural electrification programs sound rather conventional: extend electricity first to the most developed villages with the greatest pent-up demand for electricity. But the real challenge is to identify the less developed villages or regions where electricity, in conjunction with other rural development programs, provides a development stimulus. Conventional project appraisals typically ignore doing the necessary background research for assessing community responses to electrification. One exception was Thailand, where an extensive monitoring and evaluation system was developed during the early stages of its rural electrification program to understand electrification's role in rural socioeconomic development (Tuntivate and Barnes 2007). Those studies helped maintain the profitability of rural electrification schemes by minimizing the cost of electricity lines through targeting regions favorable to productive impact of electricity and conducive for the development of village load.

Electricity can be beneficial to poor villages and, under the right conditions, will not cause a financial drag on the electricity companies. Those villages with dense population centers can generate quite a bit of revenue for the electricity companies due to low investment costs. In villages situated in arid regions with significant irrigation potential, electrification can generate substantial economic benefits, especially for those nearby the distribution network. Credit and education are complementary programs that would improve the impact of electrification in poor villages and thus improve the revenue flows for the power companies. However, not all villages can develop into market towns; many will remain essentially agricultural in character, and this should be recognized during distribution planning. In short, electricity companies should try to increase electricity use in rural areas, improve their revenues, and develop incentives to maintain a high level of service reliability.

Villages distant from the central grid with poor load prospects are better served with decentralized energy programs. Extending costly electricity distribution facilities to villages should be avoided where it will not result in a significant number of new connections. In many areas, solar home systems or community electricity networks powered by biogas, diesel, wind or hydropower would be a less costly alternative than central grid electricity. For smaller villages, minimum load requirements might be established before they are connected to the central grid. Energy planning should not ignore such villages just because they may be uneconomical for connection with the central grid. Instead, decentralized programs can satisfy their rather low energy requirements to meet their more pressing energy needs.

The financial viability of rural electrification is sensitive to costs, load growth and electricity pricing. Village-selection procedures should figure more prominently in economic appraisal reports. Some programs may even require wealthier villages to defray some of the initial capital costs of providing transformers and high-tension lines to the other villages. Some villages may have rural development funds available to reduce some of the initial costs involved in extending electricity to their communities. This was the case in Mexico, where communities used infrastructure development funds to pay the electricity company for most of the cost of electricity extension (Gutierrez-Poucel 2007). If the price of electricity reflects the long-run marginal cost, this would help utilities maintain the financial resources necessary to provide reliable service. Finally, the careful placement of transformers and laying of high- and low-tension lines can help to minimize the costs of rural electrification.

Governments are very influential in determining the outcome of rural electrification programs. They create incentives and disincentives by setting prices and establishing priority categories of consumers. Many past programs became overly concerned with either residential or agricultural uses of electricity. In many programs, the productive uses of electricity garnered little attention.

The advantage of electricity is that it can be used for a variety of purposes and can have productive, residential and commercial impacts (Cabraal, Barnes, and Agarwal 2005). Not all villages will have need for all end uses, and programs must be flexible enough to adapt to local conditions and encourage electricity's multiple uses whenever possible. Electricity from the grid has the unusual feature of being produced centrally and distributed locally, having an impact that decentralizes productivity, information and social amenities.

Rural electrification can improve economic production and enhance the quality of rural life. The social benefits are quite important, but during the initial stages, they will only be available to those with relatively high incomes. The promotion of economic activity should be a priority for most electricity programs, but only under the right conditions. A program aimed only at households will contribute little toward raising rural incomes, and one with a narrow focus on economic production will not improve the quality of rural life. Ideally, rural electrification programs should have multiple impacts, including improving rural incomes, literacy, quality of life and social well-being.

Conclusion

Rural electrification during the 1980s certainly had not lived up to the rosy picture painted by its most ardent advocates, but it had not been as dismal a project as contended by the critics. The optimistic picture painted by the advocates was blurred by benefits that never materialized, poor service resulting in brownouts and blackouts and the inability of the poor to adopt electricity. Conversely, the pessimists' negative pronouncements were not borne out by many programs' substantial socioeconomic benefits, including increased educational opportunities made possible by electric lighting and greater productivity in commerce and industry. The lessons from the 1980s for today are that providing electricity for rural development is not a simple task, and care should be taken in shaping policies for promoting electricity to the poorest and most remote populations in the world.

The recent emphasis on promoting sustainable energy for all by the United Nations and other international donor agencies has not been accompanied by specific policy strategies. The goal is to increase numbers of people with electricity, with an emphasis on sustainable sources for producing power. Unfortunately, such recommendations are not based on the present or future energy needs of rural villages or the requirements of the electricity companies that will serve the expansion of new customers. For developing rural electrification strategies, it is necessary to take a hard look at the benefits and the least-cost alternatives for providing for such energy needs. This is not an easy task and requires a view of the long-term impact of various types of rural electrification for socioeconomic development.

Because of the renewed importance of rural electrification for policy makers today, it is important to learn the lessons from the early studies that evaluated the benefits and costs of rural electrification. The early studies identified several misconceptions concerning what grid or off-grid rural electrification can or cannot accomplish. The misconceptions are still relevant for the development of electricity programs today.

Rural electrification will not play a significant role in stemming migration; in fact, the reverse may be true since more migration seems to occur in regions with electricity. Rural electrification will not act as a stimulus to development without supporting programs or favorable socioeconomic conditions. Except for results from a few studies, birth rates appear to be unaffected by rural electrification. In most rural areas, electricity is not likely to be used for cooking in the foreseeable future, although Latin America is somewhat of an exception to this rule. Finally, electricity is not always cheaper than the alternatives since, in many remote villages, the extension of electricity can be quite costly.

Although it is not a magical force, electricity does provide basic energy important for rural development. It is instrumental to expanding agricultural pumping and driveshaft power in villages close to the grid where appropriate complementary conditions are in place. Electricity and education are mutually reinforcing programs since electric lights make reading in the evening much easier. Over the long term, rural electrification becomes more equitable as increasingly more households adopt it. The number of households adopting electricity continues to grow for years after a village receives electricity, reinforcing the notion that program evaluations require a long-term perspective. Both small- and large-scale farmers appear to benefit from rural electrification programs with a significant focus on agricultural development, and electrification spurs the use of agricultural innovations. Programs that stress the multiple uses of electricity can have a broad impact on social and economic development. Electricity improves households' quality of life and makes possible new energy uses (e.g., from having such household appliances as fans and televisions) that can have a broad impact on social and economic development.

While rural electrification is not the answer to all energy problems, the changes resulting from programs are cumulative and beneficial. Electricity does bring to villages new use of energy that would not otherwise have existed. Countries with grid rural electrification programs that manage to limit costs, improve connection rates and keep prices relatively unsubsidized should create a significant increase in social and economic development.

Despite significant advocacy on the part of international donors, we still do not know the impact of alternative energy programs that deliver lower levels of electricity for specific end uses due to a lack of good evaluation research. Simply stated, today there is a paucity of serious impact evaluation research on decentralized rural electrification programs. Advocacy is fine, but understanding the actual impacts of the small, decentralized alternatives and comparing them to the benefits of grid electricity are imperative for planning how to reach the remaining populations without electricity.

Rural electrification has had a significant impact on rural life, and the degree of its impact has been influenced by government policies, regional conditions, and the financial strength and vision of companies responsible for rural electrification. Significant socioeconomic development has occurred with rural electrification, but opportunities for even greater social and productive impacts may be constrained by the lack of complementary programs or conditions.

In the last 15 years, significant developments have occurred in the measurement of the benefits of rural electrification. New techniques more accurately measure both the social and economic impacts of rural electrification. These techniques have most commonly been applied to understand the impact of grid electrification. Today with the calls for renewable technologies for providing electricity in rural areas, it is time to adapt these evaluation techniques to renewable energy as well. Thus, in the next chapter the new evaluation techniques are presented to further the understanding of both grid and off-grid rural electrification impacts in the 21st century.

9. New Approaches in the 21st Century

The call by the United Nations and other international donors for *Sustainable Energy for All* is a welcome initiative befitting the 21st century. That well over 1 billion people on this planet are still without electricity seems almost inconceivable. That more than 2 billion cook on smoky fires is equally unimaginable. Recognition of these serious development problems is laudable, but the new calls involving clean energy for all seem unaware of some past lessons. While the pendulum appears to have swung once again toward optimism, the new advocates view modern energy as a cure-all for what ails developing countries.

This new optimism has not been accompanied by calls for more research on the impact of rural electrification for development. Over the past 25 years, research on rural electrification's impact has been slow but steady. Today's research results continue to echo many past findings on the costs and benefits of rural electrification, including that current optimism should not cloud the difficulties of providing modern energy to mostly poor rural populations. The contemporary research confirms that rural electrification—whether from renewable or non-renewable energy sources—works better in conjunction with complementary programs for rural development.

This research update covers both rural electrification assessment methodologies and new findings on the development impact of rural electrification. The first section focuses on the evolution of methods used to evaluate rural electrification, while the second reviews findings on its impact. This summary is not a review of all recent research on the subject; rather, it is a compilation of results from selected studies. The structure of the 1988 study is used to present recent findings on the impact of rural electrification for development, including such topics as impact on agriculture and small-scale industry, along with social and equity considerations.[11]

Evolution of Methodologies

The number of people who have gained access to electricity in recent decades has been quite remarkable. In 1971, the rural electrification rate was only 12 percent in developing countries, compared to more than 60 percent today. During the 1990s, the number of people without electricity in developing countries actually grew, owing to population growth, reaching 2 billion people. Significant progress made since then has reduced the number of people without electricity access to approximately 1.3 billion (table 9.1).

The methodologies for evaluating rural electrification also changed, reflecting the history of supplying electricity to rural areas. From the 1950s through the 1970s, rural electrification was the exclusive domain of large public electricity companies. In fact, the 1988 study of rural electrification in India, Colombia and Indonesia evaluated only programs of large public companies. During the late 1980s and 1990s, all this began to change. The public electricity companies, previously the exclusive providers of electricity to rural areas, began to face competition from providers of renewable energy and decentralized alternatives. In addition, donors became distrustful of politically controlled large public companies that often misdirected electricity to constituents at below the cost of service. The result was that, in some countries, the utilities became financially unstable and underinvested in operation and maintenance. This, in

---

[11] The findings in this chapter are based on Barnes, Samad and Banerji 2014.

turn, meant that consumers eventually experienced low levels of electricity service, including many brownouts and blackouts.

Table 9.1 Rural Electrification in Developing Countries, 2010

Region	Population without electricity (millions) 2010	Urban electricity (%) 2010	Rural electricity (%) 2010	Rural electricity (%) 1970
Africa	585	60	14	4
East Asia	182	96	86	20
South Asia	493	90	60	12
Latin America	31	99	74	23
Middle East	21	99	72	--
Developing countries	1,314	90	63	12

Source: IEA 2011.

Note: East Asia refers to China, and Africa refers to Sub-Saharan Africa; 1970 figures for East Asia and South Asia are estimated since they were reported together as 15 percent.

The result in the 1990s was a call for new ways of delivering electricity to both rural and urban people in developing countries, with an increasing emphasis on the private sector, renewable energy and decentralized distribution. These new service providers ranged from large private grid-electricity companies to community organizations providing electricity through small grids (figure 9.1). Also, a variety of organizations, including retailers, sold small home-energy devices that ranged from household systems to single lights. Thus, the means of selling electricity evolved from a heavy emphasis on central planning to more decentralized approaches, but the majority of new connections continued to be supplied through grid electricity systems.

During the same period, the method for measuring the socioeconomic benefits of electricity also changed, in part reflecting the new business environment. Forty or fifty years ago, the main benefits of rural electrification were measured as revenue plus avoided costs of using alternative fuels. The justification for using household revenues as rural electrification benefits was that this represented a minimum willingness to pay for electricity. The reasoning was that people who buy electricity perceive its value and make a decision to purchase electricity service. Therefore, the benefit must be, at a minimum, equal to what they pay for the service. Added to this were avoided costs of using other fuels, including those of diesel fuel for lighting, irrigation or motive power. Such approaches proved to be quite conservative ways of measuring the benefits of rural electrification because they underestimated the value of electricity for a variety of reasons. For instance, prices were often heavily subsidized and consumers might have been willing to pay much more than asked by politically influenced electricity companies.

Figure 9.1 Evolution of Project Methods for Evaluating Rural Electrification

Economic Benefit Measures

Consumer Surplus and Productivity Increases Due to Electricity

Consumer Surplus Based on Household kilolumen Hours

Consumer Surplus Based on Household kWh Use

Utility Revenues and Avoided Costs

Public Companies | Plus | Rural Electric Cooperatives | Plus | Private Companites, Microfinance Groups, NGOs, Energy Retailers and Others

| 50 | 40 | 30 | 20 | 10 | Today |

Years Ago

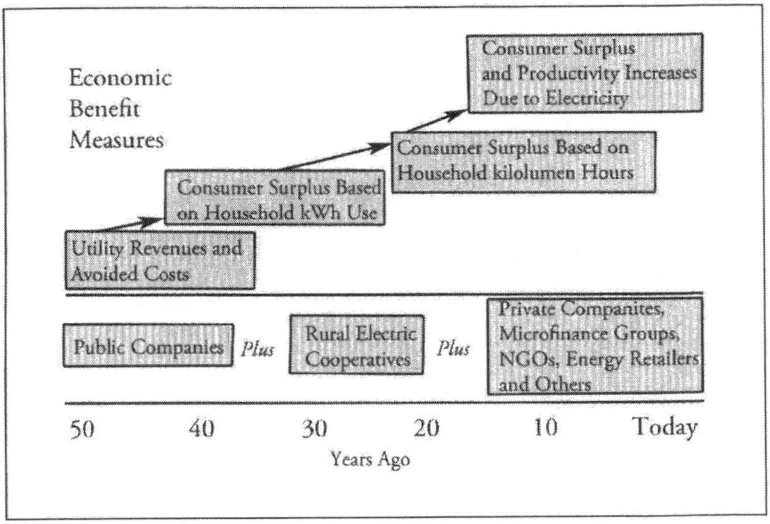

Source: This study.

During the 1990s, the willingness-to-pay methodology evolved to include consumer surplus, a relatively simple yet often misunderstood concept. Consumer surplus is the difference between what consumers are willing to pay for electricity and what they actually pay (figure 9.2, shaded area).

Figure 9.2 Consumer Surplus for Household Lighting

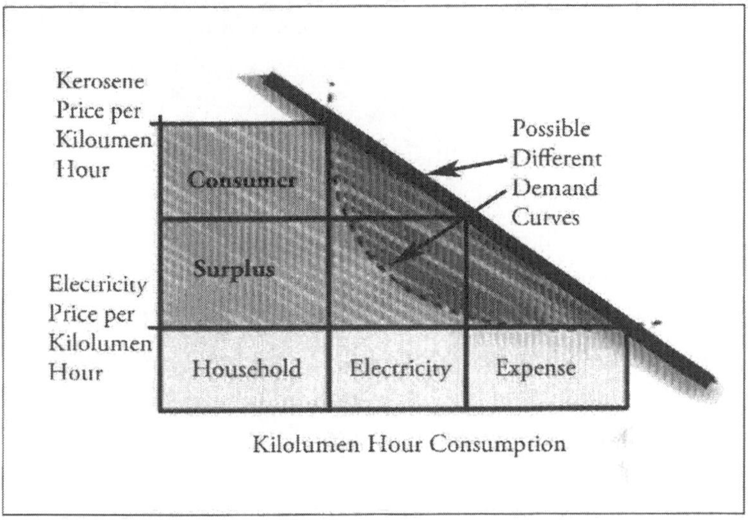

Source: Peskin 2006.
Note: This figure is for illustrative purposes; actual dimensions are dependent on pricing and consumption quantities from household surveys.

For instance, an electricity company may charge a price of 10 cents per kilowatt-hour, but a consumer who values that electricity at a much higher rate reaps a consumer surplus. During the 1980s, this idea was first applied to the demand for kilowatt-hours (Anderson 1975; Webb and Pearce 1985). This was an improvement over the use of revenues as benefits because it took into consideration consumer valuation of the benefits of electricity based on their demand. The drawback was that consumers were not actually demanding kilowatt-hours, but the activities made possible by electricity through the use of such appliances as electric lamps and machines.

The next step was to apply the method of consumer surplus to value the actual end uses of electricity. When households first adopt electricity, they use it for lighting, replacing most of their kerosene lamps. Because electric lighting is of a higher quality and much cheaper than the light produced by a kerosene lamp,[12] households can consume more light at a lower price. The result is that consumers gain the value of additional lighting by adopting electricity. A reason consumers often give for wanting more light is so that their children can study at night, which they perceive as a long-term benefit for the family. Women can cook meals more efficiently due to having light in the evening. These are the actual benefit measures, while consumer surplus for household lighting is a shortcut for indirectly measuring such benefits (figure 9.2).

As mentioned above, consumer-surplus methods for measuring benefits are often misunderstood. Those unfamiliar with the underlying theory sometimes think it is a measurement between two price-consumption points—one for electricity and one for kerosene lighting. This is wrong because those price-consumption points must be based on representative surveys that measure consumer demand for lighting; that is, a survey that can define a demand curve. Thus, the two points actually lie on a demand curve for household lighting.

The concept of consumer surplus can be applied to electricity end uses beyond lighting. Its calculation requires an estimate of demand for lighting, entertainment, communications or other services closely linked to energy. Demand is quantified as kilolumen-hours for lighting, radio listening hours and television viewing hours. Thus, it is possible to obtain a measure of consumer surplus by using demand for kilolumens, radio listening and television viewing for households using kerosene, batteries or electricity from a grid system (World Bank 2002a; O'Sullivan and Barnes 2006; Meier et al. 2010). The benefits of switching from kerosene to electricity for household lighting can be quite high, ranging from US$7 per month in the Philippines to more than $16 per month in Bolivia (World Bank 2008, p. 41). Depending on the assumptions made for consumer-demand

---

[12] It is estimated that, for the same amount of energy, electric lights produce as much as 50 times more light than kerosene lamps (van der Plas and Graaff 1988; Nieuwenhout, van de Rijt, and Wiggelinkhuizen 1998).

behavior (i.e., shape of the demand curve), the estimates can be even higher.

Whenever possible, it is better to measure the benefits of rural electrification more directly since the demand curves for other services are not easily revealed, owing to the small number of demand points (e.g., electricity, battery or kerosene appliances). In fact, the benefits from non-marketed goods and services are better measured using a variety of indirect techniques borrowed mainly from environmental cost-benefit literature.

This approach assumes that electricity is a key input to generating goods and services that directly benefit households, including education, health, entertainment and communication, comfort and protection, convenience and productivity (Tanguy 2012). This measurement requires several steps. First, it is necessary to determine a measure for each of the final outputs. For most of these, the measure is relatively straightforward. For example, education can be measured using years of schooling, entertainment by hours of watching television or listening to the radio, health by morbidity or mortality rates, convenience by time saved and productivity by output or production. Determining a measure for comfort or protection, however, may be more difficult.

The next step is to assess the differences in final outputs between households with and without electricity, which requires a carefully designed survey of representative households. The effect of electrification or change in outputs must be quantified. Because final outputs in most cases are affected by other factors (e.g., income) in addition to electrification, some type of multivariate modeling is usually needed.

Finally, quantifying the value of the impact requires estimating the households' willingness to pay for increments in final outputs resulting from electrification. The precise method used depends on the final output under consideration. For example, a household's willingness to pay for increased education could be reasonably estimated from the increase in household income resulting from this education. The income value of an additional year of education is fairly well known in most countries.

The use of consumer surplus, along with any anticipated productivity increases, can generally be used in most project appraisals, but it is important to avoid double-counting benefits. For instance, consumer surplus for household lighting estimated as a benefit may also include the benefit of household production made possible by improved lighting during the evening. The direct method is a more accurate, but less practical, way to evaluate the benefits of rural electrification. Some examples from recent research are presented in the next section.

Efficiency, Productivity and Education Gains

The relationship between rural electrification and income has been a controversial issue since the 1980s. Because wealthier households are among the first groups to adopt electricity, it is difficult to tease out the direction of causality (i.e., whether higher income causes the adoption patterns or electricity use results in higher income). Recent research has used advanced statistical models to address this issue.[13] In this section, direct estimates are used to distinguish the impact of rural electrification on income, as distinct from the well-known impact of income on electricity adoption.

Recent research in various countries confirms that adoption of electricity is a cause of higher household income (table 9.2). In Bangladesh, India and Vietnam (grid electricity) and Nepal (community-based electricity from micro-hydro) the adoption of electricity increases non-farm income.

The long-term impact of electrification on non-farm income ranges from 11 percent in Nepal up to 70 percent in India. Among the countries studied, Vietnam is the only one where adoption of electricity had an impact on farm income, perhaps because of electricity's importance to fish farming and other agricultural innovations.

---

[13] Details are provided in Khandker, Barnes, and Samad (2013a, b; Khandker et al. 2012; Banerjee, Singh, and Samad (2011); and Samad et al. (2013).

In one community in Kenya, businesses were investigated before and after being provided electricity from a local diesel generator. After these small businesses adopting electricity, their growth in gross revenues ranged between 25 percent and 100 percent (Kirubi et al. 2009), which no doubt translated into higher incomes for them.

Table 9.2 Rural Electrification Impacts on Farm and Non-Farm Household Income

Country and electricity type	Difference for households with and without electricity (%)		
	Farm income	Non-farm income	Total income
Bangladesh (grid)	31.3	35.3	21.2
India (grid)	0.0	68.8	38.6
Vietnam (grid)	0.0	27.5	28.0
Nepal (micro-hydro)	0.0	11.2	0.0
Kenya (community grid)	--	25–100	--

Sources: Khandker, Barnes, and Samad 2013a, b; Banerjee, Singh, and Samad 2011; Kirubi et al. 2009; Barnes, Khandker, and Samad 2011.

Note: Figures show percent changes in income due to household electricity access. Zero means no impact; result is not statistically significant. For Kenya, the figures are for small business revenue growth; (--) indicates data not available.

Although household electricity adoption in India appears to have no impact on farm income, this is not true for farms with electric pumps (table 9.3). A survey conducted in the late 1990s showed that, for a typical rainfed farm, the addition of an electric pump resulted in a farm income gain of more than 60 percent (approximately US $320) (Barnes, Peskin, and Fitzgerald 2002). This gain compared quite favorably to the farmers' direct electricity expenses (excluding electric pump capital expenditures), which averaged US $60 to $90 per year. For most small, medium, and large farmers with available groundwater resources, an irrigation pump appears to be a good investment.

**Table 9.3 Impact of Electric Pump Irrigation on Farm
Income in Rural India, 1996**

Farm size	Sample size (number)	Farms with electric. (%)	Farm size (hec- tare)	Income gain (%)	Income gain (US $)
Large	617	20.3	7.5	45	418
Medium	253	18.6	2.5	81	347
Small	702	13.0	1.5	54	245
Marginal	1,189	6.1	0.5	59	165
Average	2,761	12.2	2.5	68	320

Source: Barnes, Peskin, and Fitzgerald 2002.
Note: Data are from rural energy survey of six states (World Bank 2002b).

Unfortunately, irrigation investments may not have been beneficial for marginal farmers, who gained only about US $165 by adopt-
ing electric pumps. The capital costs of investing in electric pumps
and the labor involved in changing farming practices may not have
been worth the marginal benefits. Therefore, close to 95 percent of
marginal farmers did not invest in irrigation, even though it could
have led to improvement in their farm revenues. Once again, this
finding confirms that complementary enabling programs are necessary to promote the productive uses of electricity, especially for the
poorest households.

Household expenditure is a measure closely related to income.
Household income can fluctuate year to year owing to a variety of
factors. Farmers may have years of drought, workers may lose their
jobs, and business owners may have income losses during economic
downturns. By contrast, expenditures may be a more stable indicator
of household well-being because all people have monthly expenses
for food, housing and other necessities. Whenever possible, the analysis of both income (when available) and expenditures is important
to confirm the impact of electricity on household well-being.

The impact of rural electrification through large grid systems on
expenditures is quite similar to the findings for non-farm income

(table 9.4). In India, Vietnam and Bangladesh, households with grid electricity have expenditures 11–23 percent higher than those without electricity. The adoption of grid electricity also lowers the probability of households being below the poverty line by about 13 percent. For those countries with community or individual household electricity systems, the adoption of electricity has little impact on their expenditures. The amount of electricity reaching households from renewable energy systems in Nepal and Bangladesh is lower than that which is possible for the larger-scale grids, so it is quite possible that most electricity uses are for social purposes, such as socializing, reading and other activities. This still is a bit puzzling because high-quality lighting often has been found to stimulate home-business production in other countries. The findings would suggest that the decentralized renewable-energy programs could benefit from appropriately scaled complementary productive-use programs to encourage income-generating activities.

Table 9.4 Rural Electrification Impact on Expenditures and Poverty Rates

Country and electricity type	Difference for households with and without electricity (%)	
	Expenditure (% change)	Poverty (% reduction)
Bangladesh (grid)	11.3	-13.3
India (grid)	18.0	-13.3
Vietnam (grid)	22.7	--
Nepal (micro-hydro)	0.0	--
Bangladesh (SHS)	0.0	--

Sources: Khandker, Barnes, and Samad 2013a, b; Khandker et al. 2012; Banerjee, Singh, and Samad 2011; Samad et al. 2013; Barnes, Khandker, and Samad 2011.
Note: Figures show percent change in expenditure and percentage point reduction in poverty due to household access to electricity. SHS stands for solar home system. Zero stands for no impact; coefficient is not statistically significant. (--) = not estimated.

The impact of rural electrification on education was one of the most important discoveries of the 1988 study, given that it confirmed results from some earlier studies (Madigan, Herrin, and Mulcahy 1976; Saunders et al. 1978) and has been confirmed by later research (Khandker 1996). Recent research using modeling techniques to deal with the causality issue confirms the importance of both grid and decentralized electricity for education (table 9.5). Study time in the evening improved for children in households with electricity, which no doubt leads to children staying in school longer. One might have expected that having electricity would ease the household chores of girls more than boys. However, study time improved for both boys and girls, so there is no decline in the gender education gap, a finding also consistent with a recent study of rural Peru (Kulkarni and Barnes 2014). The positive outcome of decentralized electricity for education confirms the importance of even quite low levels of electricity for improving children's education. Decentralized sources of electricity may not provide the power necessary to spur productive activities, but they are valued for social benefits.

Table 9.5 Rural Electrification Impacts on Education Outcomes

Country and electricity type	Study time in the evening (minutes/day)		School enrollment (%)	
	Boys	Girls	Boys	Girls
Bangladesh (grid)	21.9	12.3	--	--
India (grid)	11.6	13.5	6.0	7.4
Vietnam (grid)	--	--	6.3	9.0
Nepal (micro-hydro)	7.7	12.0	--	--
Bangladesh (SHS)	7.0	8.2	--	--

Sources: Khandker, Barnes, and Samad 2013a, b; Banerjee, Singh, and Samad 2011; Samad et al. 2013; Khandker et al. 2012.

Note: Figures are changes in outcomes for boys and girls 5–18 years of age due to household access to electricity. SHS stands for solar home system. (--) = data not available.

One aspect of rural electrification found to be negative in the 1988 study was the equity of benefits. This persistent bias in development is a problem according to recent studies in Bangladesh, India and Vietnam. Wealthier households adopting electricity had more purchasing power and took greater advantage of the many benefits of rural electrification. In addition, better-off households were able to purchase time-saving appliances and machinery, increasing their productivity.

In the three countries, the percent increase in income due to adopting electricity ranges from well below 25 percent in the lowest income groups to more than 50 percent in the highest ones (table 9.6). It is encouraging that, except for Vietnam, the lower income groups with electricity still benefit from having electricity. Given the persistence of these results over the last 25 years, policy makers should pay special attention to the ability of the poorest households to afford various types of appliances and machines, especially focusing on those that would help lift them out of poverty.

Table 9.6 Rural Electrification Impacts: Improvements in Income and Expenditure by Income Group

Income group	Improvements in income from electricity (%)		
	Bangladesh	India	Vietnam
15th	12.4	25.9	0.0
25th	10.5	29.7	0.0
50th	15.1	36.1	29.8
75th	21.5	40.4	36.1
85th	23.9	45.7	40.6

Sources: Khandker, Barnes, and Samad 2013a, b; Khandker et al. 2012.
Note: Zero indicates no impact; result is not statistically signficant.

***

The evaluation of the benefits of rural electrification has come a long way since the 1980s. Both the means of providing access to electricity and the methods for evaluating its benefits have changed. However, with only minor differences, the message still is loud and clear that rural electrification has significant social and economic benefits for developing countries. Depending on the local circumstances, income, education and rural productivity all improve when a community gains access to electricity. The benefits of large-scale grid electricity seem to have an edge in terms of productive use, but all forms of electrification have significant social benefits for rural populations.

Given that the economic benefits of providing rural households electricity are quite high, why has the problem of extending electricity access for all persisted? The answer is that rural electrification is not a good short-term business for the electricity companies. Due to low prices, poor rural populations, inadequately designed subsidies and political interference, distribution companies often resist either providing rural areas electricity or extending high-quality service to rural residents. This is not for lack of interest in helping rural people; rather, it is the nature of the business model. Over the short term, rural electrification is fraught with financial problems.

But over the long term, the financial prospects of rural electrification are quite good for the utilities. As people gain income from having electricity, they will, over time, invest in new appliances and use more electricity. This means that both centralized and decentralized electricity service providers will have better markets and eventually will be able to profit from investments in rural electrification. Many countries with successful programs have eventually been able to eliminate special programs for rural electrification and integrate them into normal business practices (Barnes 2007). The economic benefits of rural electrification eventually mean that households have both the willingness and ability to pay for electricity.

The original 1988 study on the benefits of rural electrification highlighted the importance for developing countries to finance electricity access for all. The more recent research, which has used even more rigorous methods for evaluating the benefits of rural electrification, confirm the earlier findings from the 1980s and underscore the need for complementary programs. Today 1.3 billion of the poorest people in the world still have no electricity. Reaching them with high-quality service will not be easy, but the substantial benefits are worth the time and effort. Over the long term, the investments will reap substantial rewards for countries that take on the challenge of providing electricity for all, including the world's poorest populations.

Appendix: Research Methodology

Designing a methodology to measure the impact of rural electrification on development in India, Colombia and Indonesia was challenging in the early 1980s owing to their low levels of rural electrification. In all three countries, the surveys involved sampling households, communities and rural businesses. The various research techniques used to collect information included household interviews, reviews of village records, interviews with village leaders and data collection from rural industries. In order to measure the long-term impact of rural electrification, communities in the project had to have a wide range of experience with electricity. For this reason, some of the communities selected had had electricity for more than 20 years, while others had had it for less than 5 years or not at all. The sample also had to have diverse levels of development if generalizations were to be applicable to both developed and less developed communities. Similarly, in order to understand the effects of electricity on all income groups, the survey needed to include high-income households without electricity and poor households with electricity, as well as wealthy households with electricity and poor households without electricity. This ruled out a strictly random sample procedure.

Most past studies on the impact of rural electrification on development have concentrated on rural households as the unit of analysis. One difficulty encountered in these studies has been that wealthy households are more likely to adopt electricity compared to poor ones. As a result, income and electricity adoption are highly correlated, and this relationship may have caused spurious correlations between electricity adoption and the other socioeconomic variables studied. In order to control for the bias that would occur in a random sample, this study established sampling quotas for the India and Colombia cases. The household quotas based on occupation included rich and poor households with and without electricity through a sample stratified by occupation. Controlling for occupation reduced the possibility of spurious correlation between electrification and a household's income and between electrification and favorable characteristics associated with income, such as higher educational or lower fertility levels.

India

In India, the study villages were chosen using a combination of purposive and random selection procedures. The communities selected had varying histories of electrification, ranging from long to short to none at all. Likewise, the regions were selected for varying levels of development. The India samples exhibited a wide range of time with electricity, from no electricity to having it to over 20 years, and level of socioeconomic development. Eighty percent of the sample villages had electricity, and the average time of having had electrification was 12.6 years (table A.1).

Sampling Techniques

For India, the study selected 132 villages in 12 districts across 4 states. The selected states—Andhra Pradesh, Maharashtra, Punjab, and West Bengal—located in southern, western, northern and eastern regions, respectively, were representative of India's diverse social, cultural, and geoclimatic conditions and level of socioeconomic development. Together they accounted for 27 percent of the country's population and 22 percent of its geographical area. The selection of Andhra Pradesh, Maharashtra and West Bengal was dictated by the decision to replicate many of the questions in a 1966 study on agricultural innovations in India's villages (Fliegel et al. 1968). The sample was broadened by adding the state of Punjab, a wheat-growing region with a high level of agricultural development and rural electrification. The sampling of the villages in Punjab followed the procedure used in the 1966 study in order to avoid any bias that might result from altering sampling designs.

Table A.1 Electricity and Development Level in India Village Sample, 1980

Year of electrification	Village sample level of development (no. of villages)			
	High	Med.	Low	Total
No electricity	16	13	29	29
1975–85	7	11	17	17
1965–74	25	24	49	49
Before 1965	19	18	37	37
Total	67	65	132	132

Source: India Survey 1981.

The decision to resurvey the villages in the 1966 study made it possible to conduct an over-time or longitudinal analysis (Barnes and Binswanger 1986). Previously, most rural electrification studies had conducted cross-sectional analyses. The resurvey of the 1966 villages improved the study's ability to deal with the crucial question of whether rural electrification was the cause or an effect of important changes in village socioeconomic development. This research offered a significant opportunity to examine changes in India's rural villages over a 15-year time period. During 1966–81, 57 percent of villages gained access to electricity for the first time. The 132 sample villages were quite representative of India, although they were somewhat above average in population size and level of development. Also, the percentage of villages with electricity in the sample was higher than the national average.

Table A.2 Number of Interviews in India, 1981

Survey type	Number
Village information	132
Village leader	518
Electricity board	90
Household	631
Industry/commerce	134
Total	1,505

Source: India Survey 1981.

Surveys and Information Gathering

The fieldwork in India was conducted during March–May 1981 for Andhra Pradesh, Maharashtra and Punjab and in April–June 1981 for West Bengal. All fieldwork was coordinated by Indian researchers. Four teams of experienced economic investigators, each guided by senior field supervisors of the Operations Research Group, an India research team, conducted the fieldwork in the four states. The actual interviews commenced only after the quality of the work and the inter-team consistency of pilot surveys had met project standards. The field teams received extensive printed instructions and maps. In addition, the investigators were provided basic data from the 1971 census so that broad cross-checks with certain critical indicators were possible.

Village-level information for India was obtained by examining village or community records, through structured interviews with village leaders, and from data collected by the utilities (table A.2). The questions asked covered a wide range of topics, ranging from crop yields to the number of young men migrating from the community. The village leader survey provided fairly reliable information on aspects of the community not published in the official statistics, and in many instances, the leaders' responses acted as an independent check on the reliability of the official data. In West Bengal, for example, the data obtained from the State Electricity

Board was deemed unreliable since it did not correspond with data collected from the village survey.

In India, village-level information was obtained using three research schedules: village information, village leader, and electricity board. Much of the data for the village information schedule was acquired from the village block or official records, and other data were obtained from such sources as the village Panchayat president, village-level workers and state electricity board (table A.2).

Respondents were selected based on their leadership roles in the community. For the village leader survey, four types of leaders were selected: (i) village political leaders, (ii) village cooperative leaders, (iii) school teachers and (iv) officials of voluntary agencies (Samanta and Sundaram 1983). If one of these officials was unavailable, the alternative interviewee was selected from a formal list of alternative leaders.

Despite the differences, the occupational distributions of the samples were similar. The household survey was administered in 36 villages with electricity (9 from each state) out of the total sample of 132 villages. Ideally, the household interviews should have been conducted in control villages without electricity. But previous studies on rural electrification indicated that, in most cases, the differences between households without electricity in communities with and without access were quite small (Saunders et al. 1978). The India household sample thus consisted of a stratified random sample of households in villages with electricity. The stratification procedure was to select households with and without electricity in similar occupational categories. Consistent with the goals of the research design, the households without electricity in the more prosperous occupations and the households with electricity in the lower-paid occupations were somewhat over-sampled compared to a normal population distribution.

The industry survey was a separate research schedule. In the India survey, industry was defined as all non-household manufacturing, including both food-processing industries, such as grain mills, and manufacturing units creating a product for sale. The industry questionnaire was to be administered to all non-household manufacturing establishments in the total 132 villages, whether or not the villages had electricity. The total sample for the industrial schedules was 134 interviews or 60 percent of the total manufacturing firms in all villages. The respondent for the industrial survey was the owner or senior partner of the manufacturing firm.

Colombia

The Colombia surveys assumed that benefits for the community, households and industries could differ substantially. As a result, different types of surveys were administered in the communities to measure the various benefits. The survey included a variety of research instruments to collect information on households, industries and agriculture. The study used a combination of purposive and random selection procedures to choose communities, who had had electricity for an average of 11.4 years.

Sampling Techniques

The Colombia study followed a sampling strategy similar to the procedure used in the India study. Sixty Colombian communities were selected from three regions. The coffee-growing region was the most advanced in terms of level of socioeconomic development, while the North Coast was the least developed; however, the variance of municipalities within the zones was large. Level of income, percent literate, commercial energy use and water services formed a scale to measure a municipality's general level of development. Based on this scale, the municipalities were purposefully chosen to represent a broad range of socioeconomic development (table A.3).

Surveys and Information Gathering

The fieldwork in Colombia was completed by the research firm Instituto SER during June–July 1981. Three teams of interviewers were assigned to each region. In total, 15 persons were involved in the fieldwork, including 1 supervisor and 4 field investigators per region, all of whom were from Colombia. The interviewers received an introductory course from Instituto SER covering the data collection instruments and contents of each research schedule. Next, pilot interviews were conducted to familiarize the staff with the research instruments. In the process of these trial interviews, each interviewer was given an instruction manual containing detailed rules regarding the expectations of and materials required by Instituto SER. The supervisors and interviewers were professionals, including psychologists, sociologists, economists and social workers, and had extensive experience in this type of work.

Table A.3 Electricity and Development Level in Colombia Sample, 1981

	Village sample level of development (no. of villages)			
Year of electrification	High	Med.	Low	Total
No electricity	7	9	16	32
1975–85	3	5	8	16
1965–74	6	7	13	26
Before 1965	14	9	23	46
Total	30	30	60	120

Source: Colombia Survey 1981.

Information on rural Colombian communities was collected from household surveys, business establishments and community surveys (table A.4). The community surveys included materials from available secondary data and community leader interviews. The community leaders were those that had an affiliation with an institution, such as the mayor or priest, and those with no institutional affiliation but with considerable influence in the community, such as large landowners and civic leaders.

The community research schedule was structured so that objective characteristics of the community, such as number of businesses and schools, nearest roads and extent of community out-migration, were in a different section than the community leader's opinions on various topics. Only one leader answered questions referring to the more objective community characteristics, for it was believed that leader opinion responses would vary substantially more than those on the objective community characteristics. The opinion part of the community leader survey was administered to four village leaders. Unfortunately, attempts to collect reliable information from the power industry proved unsatisfactory. Attempts at independent verification of the power industry statistics in the 60 villages revealed that the data were completely unreliable.

Table A.4 Number of Interviews in Colombia, 1981

Survey type	Number
Community information	60
Community leader	211
Electricity company	--
Household	608
Industry/commerce	136
Total	1,015

Source: Colombia Survey 1981.
Note: (--) indicates data not available.

In the Colombia study, household interviews were conducted in all 60 rural communities. A random sampling procedure for typical occupations within communities satisfied most of the occupational and electrification quotas. In contrast to the India study, the Colombia control group of households without electricity was mainly from the communities without access to electricity. The occupational characteristics of the Colombia household sample were remarkably similar to the distribution in the India sample. The total number of households interviewed was 631 or about 10.5 interviews per community.

The Colombia survey of businesses was limited to those who were interviewed in the household survey. Special quotas were established to include household industries and shops, businesses or industries outside of the homes in the survey. A total of 136 businesses were surveyed in the 60 villages. In addition, information on the number of rural industries in the community was documented in the community survey.

Indonesia

The Indonesia survey of households and businesses with and without electricity was based on a random sample. However, 94 percent of the business population had electricity, and all were included in the analysis (table A.5). Data on home-based enterprises were gathered in the course of the household survey. Also, interviews were conducted with officials of the National Electric Company (PLN), government and bank officials and village leaders. Participant observation of the villages was conducted for five months.

Table A.5 Sample Populations in Klaten Region of Indonesia, 1981

Sample category	Total population	Sampled population
Households		
With electricity	1,748	217
Without electricity	484	118
Total	2,232	335
Businesses		
With electricity	123	123
Without electricity	8	8
Total	131	131

Source: Indonesia Survey 1981.
Note: Electricity had only been available 1–3 years.

In the Indonesia study, a series of surveys was conducted during October 1980–February 1981 in Central Java. The survey area consisted of villages where electricity had been made available for one-and-a-half years. Survey instruments were designed to obtain quantitative information on impacts directly attributable to electrification.

Off-farm, income-generating activities were examined to assess productive uses of electricity.

The Indonesia survey differed somewhat from the India and Colombia studies. Data were collected in four ways: village survey, interviews with PLN officials, interviews with village leaders and interviews with government officials at district and subdistrict levels. The village survey was conducted only in the areas of eight villages where electricity had been made available by the Klaten model rural electrification project. The village survey was in fact a series of surveys of households and businesses.

A census of all households was first conducted to obtain an initial estimate of household income. Based on those data, households with and without electricity from a stratified random sample were interviewed. Also, a census of the 131 businesses identified during the course of the survey was conducted. Information on home-based enterprises was collected as part of the household survey.

References

Administrative Staff College of India. 1980. "A Study of Costs and Benefits of Rural Electrification in Andhra Pradesh." Hyderabad study sponsored by the Rural Electrification Corporation, New Delhi, India.

Anderson, Dennis. 1975. *Costs and Benefits of Rural Electrification: A Case Study in El Salvador.* World Bank Public Utilities Report No. 5. Washington, DC: World Bank.

Asaduzzaman, M., D. Barnes, and S. Khandker. 2009. *Restoring Balance: Bangladesh's Rural Energy Realities.* Energy Sector Management Assistance Program, Special Report 006/09. Washington, DC: World Bank.

Attwood, D. W. 1979. "Why Some of the Rich Get Richer: Economic Change and Mobility in Rural Western India." *Current Anthropology* 20 (September): 495–516.

Banerjee, S., A. Singh, and H. Samad. 2011. *Power and People: The Benefits of Renewable Energy in Nepal.* Washington DC: World Bank.

Barakat, A., M. Rahman, S. Zaman, A. Podder, S. Halim, N. Ratna, M. Majid, A. Maksud, A. Karim, and S. Islam. 2002. *Economic and Social Impact Evaluation Study of the Rural Electrification Program in Bangladesh.* Report submitted to the National Rural Electric Cooperative Association International, Dhaka.

Barnes, Douglas. 1983. "Agricultural Development and Declining Women's Labor Force Participation." World Bank background paper, Washington, DC.

———. 1988. *Electric Power for Rural Growth: How Electrification Affects Rural Life in Developing Countries.* Boulder, CO: Westview Press.

———. 2007. *The Challenge of Rural Electrification: Strategies for Developing Countries.* Washington DC: RFF Press.

———. 2010. "What Is Rural Electrification? New Technologies and Old Definitions." Energy for Development, Washington, DC.

Barnes, Douglas, Frederick C. Fliegel, and Reeve D. Vanneman. 1982. "Rural Literacy and Agricultural Development: Cause or Effect?" *Rural Sociology* 47(2): 251–71.

Barnes, Douglas, and Reeve Vanneman. 1983. "Agricultural Development and Rural Landlessness in India." *Studies in Comparative International Development* 18 (Spring/Summer): 90–112.

Barnes, Douglas, and Karl G. Jechoutek. 1984. "Rural Electrification Issues: Growth, Options, and Impact." World Bank background paper, Washington DC.

Barnes, Douglas, and Hans Binswanger. 1986. "The Impact of Rural Electrification and Infrastructure on Agricultural Changes, 1966–1980." *Economic and Political Weekly* 21(1), January 4.

Barnes, D., H. Peskin, and K. Fitzgerald. 2002. "The Benefits of Rural Electrification in India: Implications for Education, Household Lighting, and Irrigation." Background paper prepared for South Asia Energy and Infrastructure, World Bank, Washington, DC.

Barnes, D., S. Khandker, and H. Samad. 2011. "Energy Poverty in Rural Bangladesh," *Energy Policy* 39: 894–904.

Barnes, D., H. Samad and S. Banerjee. 2014. "The Development Impact of Energy Access." In *Energy Poverty: Global Challenges and Local Solutions,* eds. Antoine Halff, Benjamin K. Sovacool, and Jon Rozhon. New York: Oxford University Press.

Bhatia, Ramesh. 1979. "Energy Alternatives for Irrigation Pumping: Some Results for Small Farms in North Bihar." Background paper, Institute of Economic Growth, Hyderabad, India.

———. 1983. *Investments in Infrastructure for Rural Energy: A Review of Available Studies for India,* 2 vols. Manila: Asian Development Bank.

Boserup, Ester. 1970. *Women's Role in Economic Development.* New York: St. Martins Press.

Bradford, Ernest. 1925. "The Influence of Cheap Power on Factory Location and on Farming." *The Annals of the American Academy of Political and Social Science* 118 (March): 91–95.

Brodman, Janice. 1982. "Rural Electrification and the Commercial Sector in Indonesia." Discussion Paper D-73L. Washington, DC: Resources for the Future and U.S. Agency for International Development.

Brown, .Dorris. 1971. *Agricultural Development in India's Districts.* Cambridge, MA: Harvard University Press.

Butler, Edward, Karen Poe, and Judith Tendler. 1980. *Bolivia: Rural Electrification Project Impact Evaluation.* Washington, DC: U.S. Agency for International Development.

Cabraal, R., D. Barnes, and S. Agarwal. 2005. "Productive Uses of Energy for Rural Development." *Annual Review of Environment and Resources* 30: 117–44.

Cecelski, Elizabeth, with Sandra Glatt. 1982. "The Role of Rural Electrification in Development." Discussion Paper D73E. Washington, DC: Resources for the Future.

Center for Studies in Decentralized Industries. 1979. "Impact of Rural Electrification on Decentralized Rural Industries, Kulaba District." Bombay, India, August.

———. 1980. "Impact of Rural Electrification on Decentralised Rural Industries, Ratnagiri District." Bombay, India, January.

Chinn, Dennis L. 1979. "Rural Poverty and the Structure of Farm Household Income in Developing Countries: Evidence from Taiwan." *Economic Development and Cultural Change* 22 (January): 283–301.

Costas, Philip. 1982. *Report on the Philippines Rural Electrification Impact Survey, 1981.* Report prepared for the National Rural Electric Cooperative Association, Washington, DC.

Dandekar, V. M., and Nilakantha Rath. 1971. "Policies for Equitable Distribution." In *Poverty in India*, pp. 53–111. Bombay: India School of Political Economy.

Denton, Frank H. 1979. *Lighting Up the Countryside: The Story of Electric Cooperatives in the Philippines.* Development Academy of the Philippines. Manila: The Philippines Academy Press.

Development Alternatives. 1977. "An Evaluation of the Program Performance of the International Program Division of the National Rural Electric Cooperative Association (NRECA)." Report prepared for the U.S. Agency for International Development, Washington, DC.

Dinkelman, T. 2011. "The Effects of Rural Electrification on Employment: New Evidence from South Africa." *American Economic Review* 101(7): 3078–108.

Epstein, T. Scarlet. 1973. *South India: Yesterday, Today, and Tomorrow.* New York: Holmes and Meyer.

Flavin, Christopher. 1986. "Electricity for a Developing World: New Directions." Worldwatch Paper No. 70. Washington, DC: Worldwatch Institute.

Fliegel, Frederick, Prodipto Roy, Lalit Sen, and Joseph Kivlin. 1968. *Agricultural Innovations in Indian Villages.* Hyderabad: National Institute of Community Development.

Geertz, Clifford. 1963. *Agricultural Innovation: The Process of Ecological Change.* Berkeley, CA: University of California Press.

Goddard, Paul, Gustavo Gomez, Polly Harrison, and George Hoover. 1981. *The Product Is Progress: Rural Electrification in Costa Rica.* Project Impact Evaluation No. 22. Washington, DC: U.S. Agency for International Development.

Grogan, L., and A. Sadanand. 2012. "Rural Electrification and Employment in Poor Countries: Evidence from Nicaragua." *World Development* 43: 252–65.

Gutierrez-Poucel, Luis. 2007. "From Central Planning to Decentralized Electricity Distribution in Mexico." In *The Challenge of Rural Electrification: Strategies for Developing Countries,* ed. D. Barnes, chapter 6. Washington, DC: RFF Press.

Hayami, Juyiro, and Vernon W. Ruttan. 1971. *Agricultural Development: An International Perspective.* Baltimore, MD: Johns Hopkins University Press.

Herrin, Alexandro. 1979. "Rural Electrification and Fertility Change in the Southern Philippines." *Population and Development Review* 5 (March): 61–87.

IDB (Inter-American Development Bank). 1979a. "Evaluation Report on Rural Electrification and Energy." Washington, DC, June.

———. 1979b. "Evaluation Report on Rural Development and IDB Multi-Sector Lending." Washington, DC, November.IEA (International Energy Agency). 2011. *World Energy Outlook.* Paris: International Energy Agency.

Jain, O. P. 1975. *Rural Industrialization: India's Experience and Programme for Developing Countries.* New Delhi: Commercial Publications Bureau.

Kessler, John, Janet Balantyne, Robert Maushammer, and Nelson R. Simancas. 1981. *Ecuador: Rural Electrification.* Project Impact Evaluation. Washington, DC: U.S. Agency for International Development.

Khandker, S. 1996. "Education Achievements and School Efficiency in Rural Bangladesh." World Bank Discussion Paper No. 319. Washington, DC: World Bank.

Khandker, S., D. Barnes, and H. Samad. 2013a. "Welfare Impacts of Rural Electrification: A Panel Data Analysis from Vietnam." *Economic Development and Cultural Change* 61(3): 659–92.

———. 2013b. "The Welfare Impact of Rural Electrification in Bangladesh." *Energy Journal* Vol. 33, No. 1.

Khandker, S., H. Samad, A. Rubaba, and D. Barnes. 2012. "Who Benefits Most from Rural Electrification? Evidence in India." Policy Research Working Paper No. WPS 6095. Washington, DC: World Bank.

King, Dwight Y., and Peter D. Weldon. 1977. "Income Distribution and Levels of Living in Java, 1963-1970." *Economic and Political Weekly* 25 (July): 699–711.

Kirubi, C., A. Jacobson, D. Kammen, and A. Mills. 2009. "Community-Based Electricity Micro-Grids Can Contribute to Rural Development: Evidence from Kenya." *World Development* 37: 1208–21.

Kulkarni, V., and D. Barnes. 2014. "The Impact of Rural Electrification on Education in Peru." Draft paper, Energy for Development, Washington, DC.

Ladejinsky, Wolf. 1973. "Agrarian Reform a la Punjab." In *Agrarian Reform as Unfinished Business: The Selected Papers of Wolf Ladejinsky*, ed. Louis J. Walinsky. Washington, DC: International Bank for Reconstruction and Development.

Lay, James, and Joan H. Hood. 1979. "Interim Evaluation Report: Rural Electric Cooperative of Guanacaste, R. L. and Rural Electric Cooperative of San Carlos, R. L." International Program Division, National Rural Electric Cooperative Association, Washington, DC, March.

Lerner, Daniel. 1958. *The Passing of Traditional Society: Modernizing the Middle East.* New York: Free Press.

Lipton, Michael. 1980. "Migration from Rural Areas of Poor Countries: The Impact of Rural Productivity and Income Distribution." *World Development* 8 (January): 1–24.

Lovins, Amory B. 1977. *Soft Energy Paths: Toward a Durable Peace.* Cambridge, MA: Ballinger Publishing Company.

Madigan, Francis C. 1981. "Cooperative Rural Electrification, Income Distribution, Employment, and Fertility: A Case Study from the Southern Philippines." Paper contributed to the General Conference of the International Union for the Scientific Study of Population, December 9–16, Manila.

Madigan, Francis C., Alejandro N. Herrin, and William F. Mulcahy. 1976. "An Evaluative Study of the MISAMIS Oriental Rural Electric Service Cooperative, Inc. (MORESCO)." Report prepared for the U.S. Agency for International Development, March.

Mandel, David H., Peter F. Allgeier, Gary Wasserman, Gerald Hickey, Robert Salazar, and Josephine Alviar. 1981. *The Philippines: Rural Electrification.* Project Impact Evaluation Report No. 15. Washington, DC: U.S. Agency for International Development.

McCawley, Peter. 1979. "Rural Electrification in Indonesia: Is It Time?" *Bulletin of Indonesian Economic Studies,* Jakarta, Indonesia.

Meier, P., V. Tuntivate, D. F. Barnes, S. V. Bogach, and D. Farchy. 2010. *Peru: National Survey of Rural Household Energy Use.* ESMAP Energy and Poverty Special Report 007/10. Washington, DC: World Bank.

Mellor, John W. 1976. *The New Economics: A Strategy for India and the Developing World.* Ithaca, NY: Cornell University Press.

Monari, L., and D. Mostefai. 2001. "India: Power Supply to Agriculture Summary Report." Energy Sector Unit, South Asia Regional Office. New Delhi: World Bank.

Moon, Gilbert, and NRECA (National Rural Electrification Cooperative Association). 1974. *A Report on Rural Electrification Costs, Benefits, Usages, Issues, and Developments in Five Countries.* Report prepared for the World Bank, IBRD, Washington, DC.

Muller, Frederick William. 1944. *Public Rural Electrification.* Washington, DC: American Council on Public Affairs.

Munasinghe, Mohan, and Jeremy J. Warford. 1982. *Electricity Pricing: Theory and Case Studies.* Baltimore, MD: Johns Hopkins University Press.

NCAER (National Council of Applied Economic Research). 1967. *Impact of Rural Electrification in Punjab.* New Delhi: National Council of Applied Economic Research.

———. 1970. *Economics of Rural Electrification in Kerala.* New Delhi: National Council of Applied Economic Research.

———. 1981. "Report on Rural Energy Consumption in Northern India." Environmental Research Committee, Department of Science and Technology, Government of India, New Delhi.

NEA (National Electrification Administration). 1978. "Nationwide Survey on Socio-Economic Impact of Rural Electrification." Report prepared for the U.S. Agency for International Development, Quezon, Philippines, June.

Niblock, Thomas. 1982. *Retrospective Analysis of NRECA's Activities with AID Funding of 1972-1981.* Washington, DC: National Rural Electrification Cooperative Association.

Nieuwenhout, F., P. van de Rijt, and E. Wiggelinkhuizen. 1998. *Rural Lighting Services.* Petten: Netherlands Energy Research Foundation.

NRECA (National Rural Electrification Cooperative Association). 1976. "Rural Electrification for Indonesia." Report of the NRECA Study Team, Jakarta, Indonesia, May.

ORG (Operations Research Group). 1977. "Consumer Response to Rural Electrification." Report prepared for the Rural Electrification Corporation of India, Ltd., Baroda, October.

O'Sullivan, K., and D. Barnes. 2006. *Energy Policies and Multitopic Surveys: Guidelines for Questionnaire Design in Living Standards Measurement Studies.* World Bank Working Paper No. 90. Washington, DC: World Bank.

Pendse, D. R. 1980a. "Energy Crisis and Its Impact on Energy Consumers in Third World." *Economic and Political Weekly* 15 (January 19): 107–16.

———. 1980b. "Energy Crisis and Its Impact on Energy Consumers in Third World." *Economic and Political Weekly* 15 (January 26): 175–84.

PEO (Planning Evaluation Organization). 1965. *Report on the Evaluation of Rural Electrification Programmes.* New Delhi: Planning Commission, Government of India.

Peskin, H. 2006. A Primer on Consumer Surplus Supply and Demand: Common Questions and Answers. ESMAP Knowledge Exchange Series Paper No. 5, World Bank, Washington, DC.

Robert Nathan Associates. 1979. "Contribution of AID Documentation to the Evaluation of Its Rural Electrification Projects." Report prepared for the U.S. Agency for International Development, Washington, DC, September.

Ross, James E. 1972. *Cooperative Rural Electrification: Case Studies of Pilot Projects in Latin America.* New York: Praeger Press.

Rural Electrification Panel-Committee on Power. 1979 "Sixth 5-Year Plan 1980-85: A Framework." Planning Commission, Government of India, New Delhi.

Samad, H., S. Khandker, M. Asaduzzaman, and M. Yunus. 2013. "The Benefits of Solar Home Systems: An Analysis from Bangladesh." Paper prepared for World Bank, Washington, DC.

Samanta, B. B., and K. K. Varma. 1980. *Impact of Rural Electrification on Employment.* Calcutta, India: Operations Research Group for the Rural Electrification Corporation of India, Ltd.

Samanta, B. B., and A. K. Sundaram. 1983. "Socioeconomic Impact of Rural Electrification in India." Discussion Paper D-73. Washington, DC: Resources for the Future.

Sambrani, Shreekant, Gunvant M. Desai, V. K. Gupta, and P. M. Shingi. 1974a. *Electrification in Rural Gujarat, Una Scheme.* Depth Study No.3. Centre for Management in Agriculture. Ahmedabad, India: Indian Institute of Management.

———. 1974b. *Electrification in Rural Gujarat, Bayad-Modasa Scheme.* Depth Study No.2. Centre for Management in Agriculture. Ahmedabad, India: Indian Institute of Management.

Sanderson, Fred H., and Shyamal Roy. 1979. *Food Trends and Prospects in India.* Washington, DC: The Brookings Institution.

Santamaria, Alejandro. 1980. *Mercado de Trabajo y Migraciones de Trabajo en dos Comarcas Colombianas.* Bogota: Universidad de los Andes, CEDE.

Saunders, John, J. Michael Davis, Galen C. Moses, and James E. Ross. 1978. *Rural Electrification and Development: Social and Economic Impact in Costa Rica and Colombia.* Boulder, CO: Westview Press.

Schultz, Theodore W. 1964. *Transforming Traditional Agriculture.* New Haven, CT: Yale University Press.

Sen, Lalit K. 1980. "Social Factors in Technology Choice: A Case Study of Alternative Energy Sources in Rural India." Development Discussion Paper No. 85. Cambridge, MA: Harvard University, Harvard Institute for International Development.

———. 1981. "Rural Electrification in Indonesia: Policy Options and Institutional Alternatives." Cambridge, MA: Harvard University, Harvard Institute for International Development.

Sen, Lalit K., Sudhir Wanmali, S. Bose, G. K. Misra, and K. S. Ramesh. 1971. *Planning Rural Growth Centers for Integrated Area Development: A Study in Miryalguda Taluka.* Hyderabad: National Institute of Community Development.

Sen Gupta, D. P. 1977. *Energy Planning for Karnataka State. Phase 1 Towards a More Rational Distribution of Electrical Energy.* Bangalore: Karnataka State Council for Science and Technology, India Institute of Science.

———. 1984. "Rural Electrification: Alternatives to Grid Extension." International Association of Energy Economists Annual Meeting, New Delhi, India.

Small Industry Extension Training Institute. 1976. "Impact of Electrification on Rural Industrial Development: A Study in Kurnool District, Andhra Pradesh." Hyderabad, India.

Smith, D. 1980. "Rural Electrification or Village Energization?" *Interciencia* 5 (March–April): 86–91.

Smith, D., D. B. Mehta, and Peter Hayes. 1983. "Report of the Regional Rural Electrification Survey to the Asian Development Bank." Asian Development Bank, Manila.

Tanguy, B. 2012. "Impact Analysis of Rural Electrification Projects in Sub-Saharan Africa." *World Bank Research Observer* 27(1): 33–51.

Tendler, Judith. 1979. *Rural Electrification: Linkages and Justifications.* Office of Evaluation, Bureau for Program and Policy Coordination. Washington, DC: U.S. Agency for International Development.

Tinker, Irene. 1976. "The Adverse Impact of Development on Women." In *Women and Development,* ed. Irene Tinker. Washington, DC: Overseas Development Council.

Tourkin, Steven, Robert Weintraub, and Michael J. Hartz. 1981. "Report on the Results and Methodology: Klaten Area Survey on Costs, Uses, Affordability, and Quality of Service of Electricity." Washington, DC: International Statistical Programs Center, U.S. Bureau of the Census.

Troppa, Gina. 2002. "The REA Lady: A Shining Example." *Illinois Country Living.* Springfield, Illinois.

Tuntivate, Voravate, and Douglas Barnes. 2007. "Public Distribution and Electricity Problem Solving in Rural Thailand." In *The Challenge of Rural Electrification: Strategies for Developing Countries,* ed. D. Barnes, chapter 5. Washington, DC: RFF Press.

Tyner, Wallace E., and John Adams. 1977. "Rural Electrification in India: Biogas versus Large Scale Power." *Asian Survey* 17 (August): 724–34.

UN (United Nations). 2012. *Sustainable Energy for All.* New York: United Nations.

USAID (U.S. Agency for International Development). 1983. *Power to the People: Rural Electrification Sector Summary Report.* Aid Program Evaluation Report No. 11. Washington, DC: U.S. Agency for International Development.

U.S. Bureau of the Census. 1981. "Philippines Rural Electrification Evaluation: Preliminary Results of the 1980 Household Survey." Washington, DC: International Statistical Programs.

van der Plas, R., and A. Graaff. 1988. "A Comparison of Lamps for Domestic Lighting in Developing Countries." World Bank Industry and Energy Department Working Paper (Energy Series) No. 6. Washington, DC: World Bank.

Velez, Eduardo, with Carlos Becerra and Alberto Carrasquilla. 1983. "Rural Electrification in Colombia." Report by Instituto SER de Investigación for Resources for the Future and U.S. Agency for International Development, Washington, DC, March.

Venkatesan, R., K. Ravi Shankar, Sunil Bassi, and R. K. Pachauri. 1983. "Some Aspects of Rural Electrification in India." Report by the Administrative Staff College of India for Resources for the Future and U.S. Agency for International Development, Washington, DC, February.

Wanmali, Sudhir. 1983. *Service Provision and Rural Development in India: A Study of Miryalguda Taluka.* Research Report 37. Washington, DC: International Food Policy Research Institute.

Webb, Michael, and David Pearce. 1985. *Economic Benefits of Power Supply.* World Bank Industry and Energy Department Working Paper No. 6. Washington, DC: World Bank.

World Bank. 1975. *Rural Electrification: A World Bank Paper.* Washington, DC: World Bank.

———. 1980. *Energy in Developing Countries.* Washington, DC: World Bank.

———. 1984. "1981 Power/Energy Data Sheets for 100 Developing Countries. "Economic Advisory Unit, Energy Department. Washington, DC: World Bank.

———. 2002a. *Rural Electrification and Development in the Philippines: Measuring the Social and Economic Benefits.* ESMAP Report No. 255/02. Washington, DC: World Bank.

———. 2002b. *Energy Strategies for Rural India: Evidence from Six States.* ESMAP Report No. 258/02. Washington, DC: World Bank.

———. 2004. *The Impact of Energy on Women's Lives in Rural India.* ESMAP Report No. 215/05. Washington, DC: World Bank.

———. 2008. *The Welfare Impact of Rural Electrification: A Reassessment of the Costs and Benefits.* Independent Evaluation Group (IEG) Impact Evaluation. Washington, DC: World Bank.

———. 2011. *Vietnam's Rural Electrification Story: State and People, Central and Local, Working Together.* Asia Sustainable and Alternative Energy Program (ASTAE). Washington, DC: World Bank.

81603049R00132

Made in the USA
Middletown, DE
26 July 2018

Printed and bound by CPI Group (UK) Ltd, Croydon, CR0 4YY

23/10/2024

01778240-0006